AF556006

Women in Sustainable Agriculture

Women in Sustainable Agriculture

C. SATAPATHY
Director
Amity Humanity Foundation
Bhubaneswar, Odisha

SABITA MISHRA
Senior Scientist (Agriculture Extension)
Directorate of Research on Women in Agriculture
P.O. Baramunda, Bhubaneswar, Odisha – 751 003

NEW INDIA PUBLISHING AGENCY
New Delhi – 110 034

NEW INDIA PUBLISHING AGENCY
101, Vikas Surya Plaza, CU Block, LSC Market
Pitam Pura, New Delhi 110 034, India
Phone: + 91 (11)27 34 17 17 Fax: + 91(11) 27 34 16 16
Email: info@nipabooks.com
Web: www.nipabooks.com

Feedback at feedbacks@nipabooks.com

ISBN: 978-93-83305-05-6

Composed and Designed NIPA

Preface

Women, sustainable agriculture and women empowerment are now important issues in our development. Women in Agriculture have been the national and international topic. Sustainable agriculture, participation of women in farming, methods to motivate farm women, incentives to attract them for profitable farming are both research and extension agenda at present. Authors have tried to touch all these important issues from research and extension points of view. We hope the book will be of great help to those who are concerned with sustainable agriculture and empowerment of women. We acknowledge the sources that we have cited in the book

C. Satapathy
Sabita Mishra

List of Tables

List of Boxes

List of Figures

Contents

CHAPTER 1

Meaning, Concept and Definition of Sustainable Agriculture

We mean sustainable of agriculture as preservation of productivity of land for a long period. This requires deviation from present methods of cultivation. The deviation needs knowledge about ecological process. At present we mean agriculture in association with terms like, high-tech, bio-tech and info-tech. Our traditional farming system or conventional farming system essentially means sustainable agriculture.

1. Definition of Sustainable Agriculture

There are numerous definitions of sustainable agriculture. Some are cited here to provide clear meaning of it.

1. Sustainable farming system is one that are capable of maintaining their productivity and usefulness to society indefinitely. It is the system that supports resource conserving, socially supportive, commercially competitive and environmentally sounds.

2. Lockerer (1988) stated that sustainable implies a time dimension and it is capacity of farming system to endure indefinitely.

3. Dictionary meaning: Sustainability refers to keeping an effort continuously and ability to last out and keep from falling.

4. Rutton (1988) The concept of sustainability should serve as guide to agriculture practices and it must also include the use of technology and practices that both sustainable and enhance productivity to meet the increasing food demand.

5. Oxford Dictionary: Sustainable is defined as the ability to maintain without deteriorating through continuous effort.

6. Sustainability encompasses the elements of productivity, profitability, conservation, health sustainable safety and the environment seeks to optimize the skill and technology to achieve long term stability in agriculture enterprise.

7. Sustainable agriculture is the way of practicing agriculture in which there are numerous definitions of sustainable agriculture.

8. Sustainable agriculture is a whole system approach to food, feed and other fiber production that balances environment soundness, social equity and economic viability among all sectors of the public including international and intergenerational.

9. Sustainable agriculture is the farming system and a practice which maintains or enhances (i) the economic viability of agriculture production (ii) maintains resource base and (iii) other ecosystem which are influenced by the agriculture production. (Sustainability=Productivity + Conservation)

10. Stinner and House (1987) the sustainable agriculture as system is typically associated with low inputs (LISA= Low Input Sustainable Agriculture)

11. Edward (1987) defined sustainable agriculture as the minimal dependence on synthetic fertilizers, pesticides and antibiotic and more dependent on use of manures, crop rotation and minimum tillage.

12. TAC (Technical Advisory Committee) (1989) of the Consultative Group on International Agriculture Research (CGIAR) defined sustainable agriculture as successful management of resources for agriculture to satisfy changing human needs while maintaining or enhancing the quality of the environment and conserving natural resources.

13. The Agriculture Research Service (U.S. Department of Agriculture) defines sustainable agriculture as that agriculture for the foreseeable future will be productive, competitive and profitable, conserve natural resources , protect the environment and enhance the public health, food quality and safety.

14. Huang (1994) defined sustainable agriculture has three basic values like: (i) Ecologically sound (ii) Economically viable and (iii) Socially just.

15. The Codex Alimentarius Commission defines organic agriculture as a holistic food production management system which promotes and enhances agro-ecosystem health including biodiversity, biological cycles and soil biological activities. It emphasizes the use of management practices in preference to the use of off-farm inputs taking into account that regional conditions require locally adaptable systems. This is accomplished by using where possible, agronomic, biological and mechanical methods as opposed to use synthetic materials to fulfill any specific function within the system. In short, the definition has wider scope to elaborate definition of sustainable agriculture. The major components of technological intervention that need attention of all concern are:

Box No.1.1. Major components of sustainable agriculture
1. Cultural practices 2. Soil and water management 3. Non-Chemical pest and weed control 4. Integrated plant animal production 5. Nutrient recycling

Implications of sustainable agriculture: Implication of sustainable agriculture is those which are directly concerned with farmers and farm women who produce food grains for society. Summarizing a number of implications of sustainable agriculture it can be inferred in five important terms.

1. Meeting basic needs
2. Keeping population density below carrying capacity
3. Renewal of renewable resources
4. Conserving, recycling of renewable resources
5. Minimize ill effects of environmental degradation

USDA (1980) expressed the concern over five important issues of agriculture that need attention are:

1. Increased cost and dependence on chemicals and energy
2. Continued decline in soil productivity
3. Contamination of surface and ground water because of fertilizers and pesticides
4. Hazards to animal health and food quality
5. Fading out family farm and local markets

The concern about increased cost involved in agriculture, more dependence on chemical fertilizers, pollution arising out of chemical use, health hazards and departure of families from farm are quite genuine. Unless these aspects are taken care of, the future of our society will be at a stake.

II. Agro-ecosystem

Sustainable agriculture and ecosystem are two sides of a coin. Agro- ecosystem is the basic foundation of agricultural production and productivity. Sustainability of agro-ecosystem is one of the top ranking agenda of our national planning. Agro-ecosystem is sustainable when we observe the following:

1. Maintain natural resource base
2. Rely on minimum artificial inputs
3. Manage pest and disease through internal regulating mechanism
4. Recovers from disturbances caused by cultivation and harvest

To give a real meaning to sustainable agriculture besides above core factors, we shall have to include essential elements like:

1. Crop rotation
2. Integrated pest management techniques
3. Increased biological and mechanical weed control
4. Use of natural inputs

Besides above practicable elements, the internationally accepted five components of sustainable agriculture are:

1. Production to meet basic needs
2. Conservation of natural resources
3. Quality of environment
4. Community and gender equality
5. Avoidance of regional imbalance

The essential components and their relationship appear in figure given.

Advantages of Sustainable Agriculture

The advantages of sustainable agriculture need no emphasis.

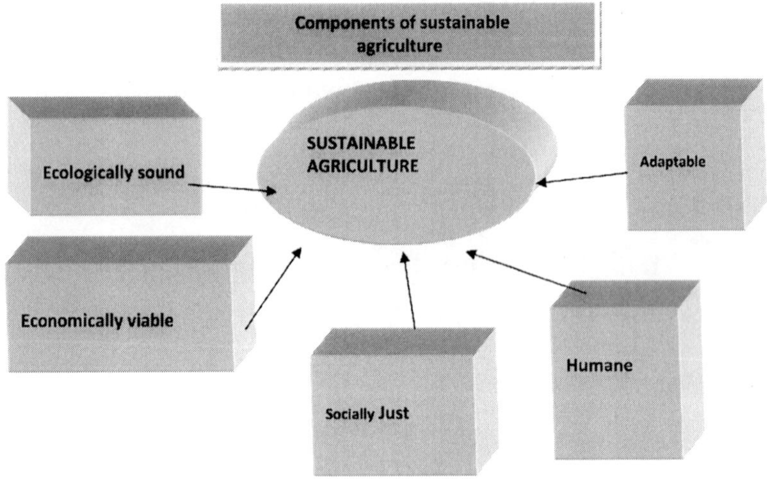

Fig. 1.1: Major Components of Sustainable Agriculture

In short the following advantages are very much visible and can be observed in real ground situation.

1. Diversity of crops and products reduces incidence of pest and diseases
2. Cost of production gets reduced
3. Crops become tolerate to stress condition
4. Sustainable agriculture is the model of social and economic change
5. Preserves bio-diversity and maintains soil
6. Uses locally available resources
7. It respects ecological principles of diversity
8. Long term good effects of agriculture

Guiding Principles of Sustainable Agriculture: Principles in agriculture and that too in case of sustainability are of prime importance to propagate the concept among the farm families.

There are cases where the guiding principle is ignored and farmers meet with loss and consequently losing faith in sustainable agriculture. The principles are cited in following boxes.

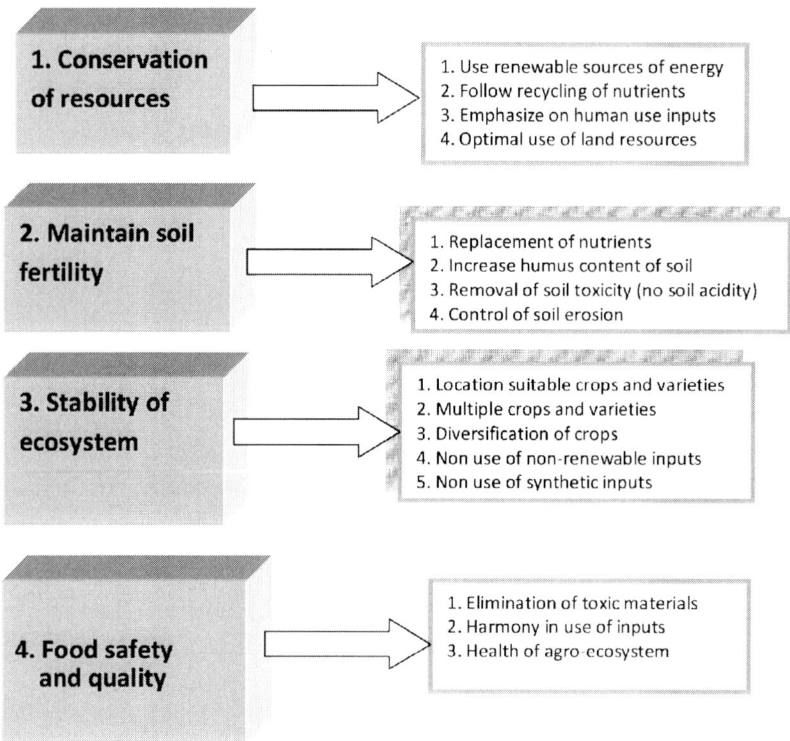

Fig. 1.2: Guiding Principles of Sustainable Agriculture

III. Approaches in Sustainable Agriculture

The concept "Sustainable Agriculture" is very broad in scope and operation. Many authors have defined and operationally stated it from different angles with some minimum and constant elements. To analyze the concept, the interchangeable terms used like: (i) Low input agriculture (ii) Biological farming

(ii) Bio-dynamic farming (iv) Organic farming and (v) Natural farming. All these terms have some difference but in totality they mean almost same content.

1. Low input Agriculture: Low input agriculture means use of locally available resources for which one does not pay much nor search here and there to find them. The low input agriculture is stable and has long term sustainability in producing required food grains. Many farm scientists apprehend whether low input agriculture can produce enough to meet the increasing demand of food grains. To have a balance in meeting food grain demands it is not desirable to discontinue present system of farming in which synthetic chemicals have greater role to play. Looking for alternative agriculture giving emphasis on farm management skill, using Nitrogen fixing legumes, growing green manure crop and adopting Integrated Pest Management system may be the right answer to the posed questions. In an opinion pool survey in Odisha, the response of farmer and farm women shows a negative trend to replace present approach in farming for non-chemical approach. Many of them do not believe on the possibility of non-synthetic farming system to meet the requirements. It is more so in case of small and marginal farmers who cannot afford for experiment even. At present, labor cost is much more than other input cost, so also cost of farm machineries which are yet to be popular.

The alternative approach for sustainable agriculture is adoption of non-chemical methods like cultural practices, mechanical and biological practices of pest control which can help to conserve resources. Some scientists are in view that less intensive farming, modified cropping sequence and crop diversification are the right answers to sustainable agriculture.

2. Biological Farming: Biological farming is the common term used in sustainable agriculture. It is known as eco-agriculture. Biological farming is a system of activating and harnessing of biological energy on a sustainable basis for the production of high quality and profitable agricultural produce. It keeps harmony with nature and a holistic approach that improves soil health which in turn improves plant, livestock and human health. It produces high quality grain and quite profitable. The consumer's demand for such agricultural produce is growing throughout the globe.

Box No. 1.2. Benefits of biological agriculture
1. Biology in soil
2. Mineralizing the soil
3. Improved soil structure

3. Bio-dynamic Farming: Biodynamic concept is that soil is living and keeps it living always. It is the combination of biological and dynamic inputs to improve soil health and living organisms in soil.

Box No. 1.3. Biological practices and Dynamic practices	
Biological practices	**Dynamic practices**
1. Green manure	1. Compost preparation
2. Cover cropping	2. Foliar spray
3. Composting	3. Planting by calendar
4. Integration of crop and livestock	4. Preparing pest control
5. Tillage and cultivation	5. Homeopathy

Concept of Bio-dynamic Farming is conceptualized to provide basic understanding of the following parameters.

1. Farm is a living organism
2. It is a self content entity
3. It has own individuality
4. Integration of crop and livestock
5. Recycling of nutrients

6. Maintenance of health of soil, crop and livestock
7. Holistic management approach

4. Organic Farming: The most common term used in sustainable agriculture is organic farming. The need and demand for organic food are in rising trend throughout the world. The concept "Organic Farming" requires fulfillment of the following characteristics.

a. Form of agriculture based on eco-system
b. Eliminate external agricultural inputs (Synthetic)
c. A holistic production management system
d. Enhance agro - ecosystem, health, bio diversity and soil biological nutrients.
e. Accounts for regional condition and adoptable system
f. Use both traditional and scientific knowledge
g. Lay emphasis on agronomy, bio-logical and mechanical methods
h. Opposes synthetic materials
i. Encourages cultural practices
j. It is a system of farm design and management to create ecosystem which can achieve sustainable productivity
k. It recognizes alternate farming system
l. Concentrates on crop rotation, residues, manures, legume, green manure and biological methods of pest control

IV. Organic Farm and Land by Continent

The area under organic farming is increasing throughout the globe. The expansion of organic farm and organic produce in India is not up to expectation. A worldwide picture of organic farm can be glanced in the table given below.

Table 1.1: Position of organic farming in world

S.N.	Continents	Organic land area (ha)	Organic farm	% of total
1.	Africa	80,504	124, 80	20.00
2.	Asia	2,893,322	120,805	20.00
3.	Europe	6,920,462	187,69	30.00
4.	Latin America	5,809,320	176,710	28.00
5.	North America	2,199,225	12,063	02.00
6.	Oceania	11,845,100	2,68	0.00
	Total	30,558,183	633,891	100.00

(*Source:* SOEL-FIBL Survey, 2007)

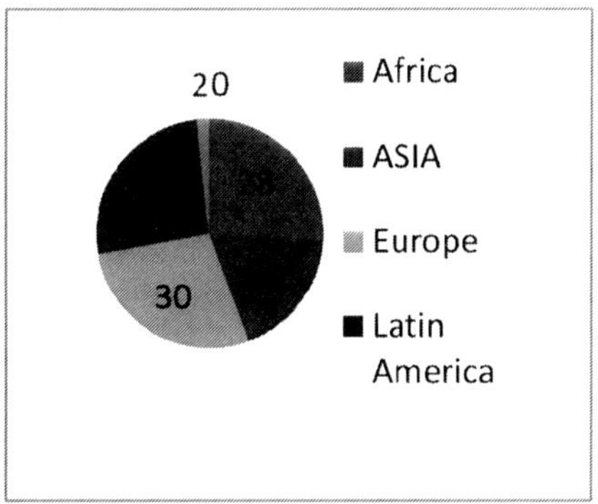

European countries are more concerned about organic farming. In India organic farming has yet to take complete shape. Organic farming is now confined to large land holding families with higher level of education and diversified occupation and mobility in nature. However, the organic farming picture of India is as follows.

Table 1.2: Growth of area under organic management

Si. No	Year	Organic area (ha)
1.	2003-04	42,000
2.	2004-05	76,000
3.	2005-06	1,73,000
4.	2006-07	5,38.000
5.	2007-08	8,65,000
6.	2008-09	12,.07,000
7.	2009-10	10,85,648

(*Source:* NCOF, 2008)

Among the states in India, Madhya Pradesh, Kerala, Assam, Rajasthan, Himachal Pradesh and Maharastra are leading in coverage of area under organic farming.

Principles of organic farming: Like other occupation and enterprise, organic farming concept is based on certain principles. These are:

a. Maintaining of living soil

b. Making available all organic essential nutrients

c. Organic mulching for conservation

d. Attaining sustainable high yield

Objectives: Objective of organic farming is to divert attention to soil and its improvement to keep productivity of crops sustainable. A brief summary of objectives of organic farming mentioned by different authors are of these kinds.

a. Produce high nutritional quality food

b. Work with harmony of natural system

c. Involves micro organism, soil, flora-fauna, plants and animals

d. Increase long term fertility of soil

e. Make use of renewable resources

f. Maintain genetic diversity

g. Ensures adequate return and satisfaction

h. Consider ecological impact

To achieve the objectives of organic farming in rural areas, certain essential elements are to be given due importance.

(i) Sustainable use of local resources

(ii) Minimal use of external inputs

(iii) Ensure biological function of soil and water

(iv) Create attractive overall landscape

(v) Increase crop and animal intensity

Major Components of organic farming:

a. Crop and soil management

b. On farm waste recycling

c. Non- chemical weed management

d. Biological pest control

Natural farming: Organic farming is synonymous to natural farming. The organic farming is not exclusive of modern farming while natural farming devoid of all components found in modern farming system. The essential principles of natural farming are, (a) no cultivation (b) no chemical (c) no weeding and (d) no plant protection measures. However, natural farming is identified by the following criteria.

a. Based on observation

b. No cultivation

c. No chemical fertilizer or prepared compost

d. No weeding by tillage or herbicides

e. No dependence on chemicals

V. Good Agricultural Practices

Now the common question in farming is good practices. Good Agricultural Practices is of recent concept which essentially indicates three important indicators.

1. Effective use of resources
2. Maximize benefits from export of agricultural produces
3. Use of sustainable agriculture

Broadly defined Good Agricultural Practice (GAP) applies recommendations and available knowledge to address environmental, economic and social issues for production and productivity. The concept of Good Agricultural Practices was evolved in recent years in the context to meet (i) food security,(ii) food quality,(iii) production efficiency, (iv) livelihoods and (v) environmental benefits at the local, national and international level.

The Good Agricultural Practices also Consider

1. Allowing acceptable level of pest damage
2. Encouraging predatory beneficial insects to control pests
3. Careful crop selection along with resistant varieties to diseases and pest
4. Planting of companion crops to divert pest
5. Use of row cover to protect crops from pest
6. Crop rotation in different localities from year to year and
7. Use of insect traps

The concept "Good Agricultural Practice"

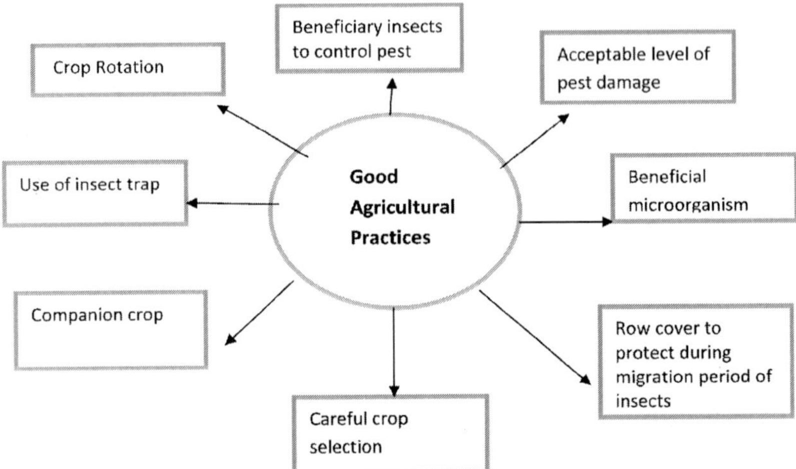

Fig. 1.3: Good Agricultural Practices

Categories of Management Practices: Best Management Practices term has been extensively used to indicate the adoption of the following practices.

Table 1.3: Management practices

Management Practices	Operations
1. Crop Management	Tillage, crop sequence, seed improvement
2. Soil and Water Management	Runoff and erosion control, moisture conservation practices, wind erosion control
3. Nutrient Management	Quantity of nutrients applied, method of application, split doses
4. Pest Management	Sanitation, pesticides usages, pest resistant crop

Parameters of Best Management Practices: Best Management Practices is a concept that has wider scope to include better practices from time to time. As such the best method of today is discarded tomorrow. For the present scenario the concept carries meaning. A farmer knows what is best to his soil, land, family and community. To keep balance between food grain requirements and soil fertility the parameters are chosen as shown in diagram.

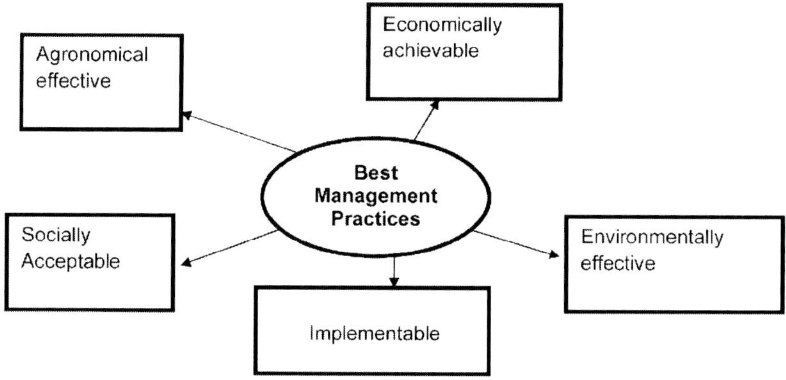

Fig. 1.4: Best Management Practices

1. **Agronomic Effectiveness:** Agronomy of crops takes care of crop and cropping system. To promote productivity and quality of commodity recent technological practices can be examined in the line of sustainability which emphasizes preservation for future.

2. **Environmental Effectiveness:** Environment friendly practices are to be screened out and practiced to save the ecology. To reduce pollution, keep atmosphere free of gas house emission, different methods are to be adopted.

3. **Economically feasible:** Market and customer demands are the top most agenda to keep farmer survive on farming. A balance sheet has to develop keeping economic benefit at one side and use of input to produce more so that sustainable agriculture could be achieved.

4. **Socially acceptable:** Compatibility of practice with existing farming system, socio-cultural traits of society and psychology of upcoming farmers would decide the nature and extent of best management practices what we conceive to run our agriculture.

5. **Implementability:** This is the degree to which we can consider the practice to put into action. The legal aspects, government program, affordability of the farmers etc. play important role in deciding fate of the best management practices that we advocate.

5. **Women and Agriculture:** In view of critical role of women in the agriculture and allied sectors, as producers, concentrated efforts were made to benefit them in terms of capacity building, training, extension services and various programs to reach them. The program for training women in soil conservation, social forestry, dairy development and other occupations allied to agriculture like horticulture, livestock including small animal, poultry, fisheries etc. were envisaged to expand benefit to women. In this regard, it is important to mention that one of the

important features of the PMNREGA is that it has turned into a magnet for women. More women than men work under national program that guarantees employment to rural people. Their participation has been growing since the inception of the Act in 2006. This is remarkable that only 28.7% women form a part of the country's work force. The toxicity and residues of the agro-chemicals in farm produce are the main problems facing mankind today. The consumers all over world are concerned about the presence of toxic residues of chemical fertilizers and pesticides in the food items including cereals, pulses, oil seeds, spices, feed fodder, vegetables, fruits, milk and milk products. The women are more concerned about health hazards arising out of inorganic farming.

Now it is time for to look alternatives. There are many potential alternatives to pesticides, chemical fertilizers and other agro-chemicals available in nature which improves and sustains the economy and natural resources. Now it is time to opt for biological and natural inputs in farming.

6. **Constraints in adoption of sustainable agricultural practice**. Organic farming management approaches has been developed for various crops/cropping systems and adequate steps have been taken by several volunteer orgainizations, government agencies, research institutions and non-governmental organizations to implement their technology. The technology does not gain popularity among the farming communities owing some socio-economic constraints.

1. Organic farming has not become popular among the farmers because of poor awareness about the utility and importance.
2. Organic farming is considered skill, knowledge intensive and needs much understanding of the organic sources of plant nutrients, bio-pesticides which becomes hindrance.

3. Application of chemical fertilizers, herbicides, insecticides, fungicides and chemical plant growth regulators are still seen as progressive approach by the farming communities. Agrochemical companies make much effort to push their products into farming system.

4. Farming community does not find this approach acceptable as they cannot change their attitude towards use of chemical fertilizers because of immediate benefits.

5. Information about organic sources of plant nutrients, bio-pesticides etc. lack to great extent for which the popularity of organic farming is low.

6. Lack of proper coordination among researchers, extension professionals and the farmers.

7. Slow action of organic inputs like FYM, green manure, compost, phosphorus, crop residue etc. take time to create confidence in the mind of the farmers.

8. Inadequate supply of bio-fertilizers, bio-pesticides, vermin compost make organic farming difficult.

9. Lack of suitable literature on organic farming and effective extension service.

10. Difficult to market produce under organic product in local areas.

11. Market for organic product is limited.

12. Shortage of easy and cheap accredited certifying agencies for organic farming and products.

Consumers in many countries willingly pay a premium price for organic farmed fruits, vegetables and other food products. Organic product enjoy 25-30% premium over non –organic products.

VI. History of sustainable agriculture

Table 1.4: History of sustainable agriculture

Si. No.	Year	Author/Organization/Government	Text
1.	300 BC	Artha Sastra	Soil management , commodity Trade
2.	400 BC	Krishi Parashar	Book on Agriculture and Soil Management
3..	1400-5500 BC	Ramayana and Mahabharata	Water Management, Kamadhenu
4.	2500-1500 BC	Vedas	Organic Manure and Green Manure
5.	590 AD	Holy Koran	Recycling in Soil
6.	1924	STAINER	Comprehensive Organic Farming System
7.	1940	Albert Howard	Father of Modern Organic Agriculture
8.	1950	USA Rodel	Sustainable Agriculture
9.	1962	Rachel	Environmental Movement
10.	1965	Mexico	High Yielding variety of Wheat
11.	1967	IRRI	High Yielding variety of Paddy
12.	1968	William Gaud	Green Revolution
13.	1970	Worldwide Movement	Know your Farmer, know your Food
14.	1972	Limit of Growth, Rome	Sustainable Agriculture Development
15.	1972	International Federation of Organic Agriculture Movement	Organic Farming
16.	1978	Bill Mobilisation, Australia	Permaculture
17.	1980	USA Global 2000	Sustainable Agriculture

Contd.

Si. No.	Year	Author/Organization/Government	Text
18.	1980	International Union of Conservation of Nature and Natural Resources	Concept of Sustainability
19.	1981	FAO Agriculture towards 2000	Sustainable Production
20.	1984	USA National Research Council	Alternative Agriculture Methods
21.	1985	Technical Advisory Committee (TAC) of CGIAR	Sustainable Agriculture
22.	1986	National Agriculture Policy	Use of Watershed Development
23.	1986	USA	Farming Act on Research of Sustainable Agriculture
24.	1986	India	Ground water Policy
25.	1986-87	India	NWDPRA
26.	1987	The World Committee on Environment and Development	Our Common future: Importance of sustainable agriculture
27.	1987	National Water Policy in India	Optimum use of water resources
28.	1988	TAC of CGIAR -Workshop	Sustainability of agriculture
29.	1988	Annual research on LEISA by USDA	Research priority on sustainable agriculture
30.	1989	TAC	Suggestions for International Agricultural Research on sustainability
31.	1989	FAO	Global agriculture, sustainable development, management of natural resources
32.	1989	NRC	Alternative Agriculture

Contd.

Si. No.	Year	Author/Organization/Government	Text
33.	1989	Board of Agriculture of the National Research Council of US	Major study published "Alternative Agriculture"
34.	1989	World Commission on Environment and Development organized seven UN conference	Environment and Development, developed most widely definition of sustainable development
35.	1990	Farming Act of US	Defined sustainable agriculture and renamed ISA to sustainable agriculture and education (SARE)
36.	1991	Establishment of World Sustainable Agriculture	Activities started in US, India, Japan, Thailand, Taiwan, Australia, Beijing
37.	1992	Earth Summit at Rio by United Nationals Conference (UNCED)	Environment and Development
38.	2000	Indian Planning Commission	Identified Organic Farming as National Challenge

VII. Agenda for Gender Equality

To reduce gender inequality, efforts are made at different levels. These are domestic and international levels. At domestic level, local governments and international level institutions are concerned for removal of inequality. The motivation for an agenda for global action is three fold, (i) Providing financial resources (ii) Fostering innovation and learning and (iii) Leveraging partnership.

1. **Providing financial support:** Improvement in the delivery of clean water and sanitation or better health services requires huge resources. The poor countries cannot afford for it for which global efforts are made to provide. The international development community can financially support countries willing and able to undertake such reforms to ensure maximum impact.
2. **Fostering innovation and learning**: The development communities promote innovation and learning through experimentation and evaluation in ways that pay attention to results, process, context in upscaling the degree of gender equality.
3. **Leveraging effective partnership:** Successful reforms require condition or partnership that can act within and across borders. Partnership requires among those in the international development communities on funding issues.

Four principles for policy and program design that can enhance the impact and effectiveness of global action across all priority areas are:

1. Comprehensive gender diagnostic as the basis for policy and program design
2. Targeting determinants versus targeting out-comes
3. Up-streaming and strategic mainstreaming
4. No size fits all

At present, there is increased difference for mainstreaming for women. The topic has broad attention at local, national and international level. The problem of women is not confined to India alone. Livelihood system, gender issues and areas of concerned are of prime importance. It is established that for improvement in agriculture and restoration of natural resources women play important roles. It is not only the contribution to agriculture in the climate change process but also impact is possible on farm women sector. Sustainable agriculture practices are directly related to gender issues and problems. Modern agriculture and need for engendering is at the top of agenda in our development. The inputs intensive in agriculture also needs the intensive efforts of women in food security, cultivation of crops, post harvesting and marketing. Role of farm women in technology adoption has been elaborated by many scholars and has been come to end that the boost of production and productivity, gender equality is of primary concern. At global level, there has been concerted effort by development of communities. Plan for rural mainstreaming as has been proposed by World Bank are discussed below.

Action plan: There are priority areas for actions to be taken up. The priority areas are classified into five important groups.

The details of global action plan for tackling priority area to reduce gender gap is presented below.

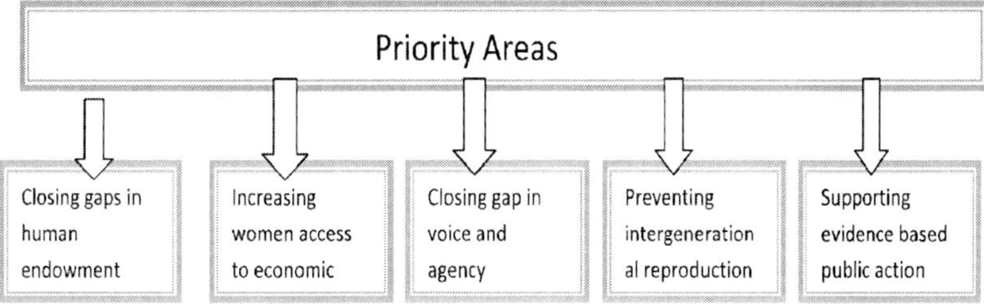

Fig. 1.5: Priority Areas

Table 1.5: Agenda for global action at a glance

Priority area	Initiatives that need support	Direction for global development communities		
		Providing financial support	Fostering innovation and learning	Leveraging partnership
Closing gender gap in human endowment	1. Increasing access to education among disadvantaged group	Yes	-	Yes
	2. Increasing access to clean water	Yes	Yes	-
	3. Strengthening support for prevention and treatment of HIV/AIDS	Yes	Yes	Yes
	4. Increasing access to specialized maternal services	Yes	Yes	Yes
Promoting women access to economic opportunities	1. Increasing access to child care and early childhood development	-	Yes	Yes
	2. Investing in rural women	-	Yes	-
Closing gender gap in voice and agency	1. Increasing women access to justice system	-	Yes	Yes
	2. Shifting norms regarding violence against women	-	Yes	-
Preventing intergenerational reproduction of gender inequality	1. Investing in adolescent girls and boys	Yes	-	Yes
Supporting evidence based public action	1. Facilitating knowledge sharing and learning	-	Yes	-

(*Source:* 2012 Gender Equality and Development, The World Bank p.351)

The strategy for removal of inequality can be analyzed as:

(i) Inequality problem at home

(ii) Inequality problem in farm sector

(iii) Inequality problem in both farm and home

The models suggested by different authors and concerned at global level infer creation of economic opportunities for improved livelihood system.

At global level it has been realized that all the countries are becoming conscious of women rights. Despite in hardships many women endure in their daily lives. Things are changed for better. In four major areas like women's rights, education, health and labour force have been remarked with change. Improvement that proves hundred years in rich countries took just forty years in some low and middle income families. Change has also been accelerative with gender equality games in every decade.

Better outcomes for women are matching in many domains. This evidence by the fact more girls are in schools, healthier life of women, more participation of women in market and new labour market opportunity is coming up in all countries of the world. It is also a fact that investment for women in all countries is increased and will be more in future.

On the basis of participation in education for children within 12 to 15 years and preventive health services the country has been categoried as follows:

Table 1.6: Inequality in educational participation of children (12 – 15 years)

	Low inequality	Moderate inequality	High inequality
1.	Kenia	Tanzania	Madagascar
2.	Uganda	Haiti	Liberia
3.	Dominician Republic	Vietnam	Nigeria
4.	Congo, Rep. (Brazzaville)	Zimbabwe	Burkina Faso
5.	Swaziland	Malawi	Ethiopia
6.	Albania	Ghana	Morocco
7.	Guyana	Peru	Senegal
8.	Jordan	Bolivia	Sierra Leone
9.	Ukraine	Colombia	Guinea
10.	Armenia	Mozambique	Mali
11.	Namibia	Cameroon	India
12.	Maldives	Egypt, Arab Rep.	Niger
13.	Zambia	Cambodia	Chad
14.	Moldova	Congo, Dem.Rep.	Benin
15.	Azerbaijan	Lesotho	Turkey
16.	-	Nepal	Cote d'lvoire

(*Source:* WDR-2012 team, estimate based on Demographic and Health Survey)

Table 1.7: Inequality in use of preventive health service

	Low inequality	Moderate inequality	High inequality
1.	Senegal	Indonesia	Nigeria
2.	Bangladesh	Malawi	Armenia
3.	Honduras	Philippines	Congo, Dem Rep.
4.	Egypt, Arab rep.	Swaziland	Niger
5.	Maldives	Colomba	Guinea
6.	Bolivia	Cameroon	Turkey
7.	Rwanda	Namibia	Congo, Rep.(Brazzaville)
8.	Sierra Leone	Kenya	Ethiopia
9.	Uganda	Peru	Mozambique
10.	Morocco	Lesotho	Benin
11.	Zambia	Domonician Rep.	Liberia
12.	Ghana	Modova	Masagascar
13.	Jordan	Burkina Faso	Zimbabwe
14.	Cambodia	Nepal	India
15.	Albania	Mali	Azerbaijan
16.	-	-	Haiti

(*Source:* WDR-2012 team, estimate based on Demographic and Health Survey)

CHAPTER 2

Women and Sustainable Agriculture

Women constitute 50% of our population. Their contribution to farming needs no emphasis as they are integral part of it. Women have developed farming in our country and contributed immensely for growth and maintenance of agriculture as a whole. But majority of them are not fully aware of sustainable agriculture in scientific terms although they practice it. Surveys and observations reveal that women are in favor of traditional agriculture and feel that traditional farming is more sustainable than the present one.

I. Work Participation of Women in Agriculture

In all states of India women participate in farming. The difference of participation is as per prevailing social system of the locality. Most of them participate in post harvest stage of farming operations. The entire animal husbandry is managed by women. In all farming operations women participate in varying degree. A study from Karnal report on work participation of women in farming operation is as follows.

Table 2.1: Gender participation in farm activities

	Farm Activities	Male Participation (%)	Female Participation (%)
1.	Land preparation	100	-
2.	Seed preparation for sowing	8	92
3.	Raising nursery	25	75
4.	Direct sowing	92	8
5.	Irrigation	83	17
6.	Applying FYM	75	25
7.	Fertilizer application	83	17
8.	Weeding	17	83
9.	Plant protection	83	17
10.	Harvesting	42	58
11.	Threshing	58	42
12.	Bagging/Storing	17	83
13.	Marketing of produce	92	8
14.	Storing dry fodder	66	34

Source: Ahmed,S. 2004.Gender Issues in Agriculture and Livelihood, Vol No.1,KAU & Swaminathan Research Foundation.

The participation of women in farming operations is found in all rural sectors. Their participation in post harvest phase is much more followed by seed and seeding along with transplanting. As reported from Karnal, their participation in land preparation is absence because of physical strength and social customs. In other parts of the country specifically in tribal areas (Koraput, Odisha) women are actively involved in ploughing and land preparation. It is seen that women participation in agriculture is comparatively more in dry land areas because of situational factors. The overall work share of women in farming is given in Box No. 2.1

Box No. 2.1: Work shared by women	
Agricultural operation	**Work share (%)**
1. Weeding	80-100
2. Soil working	50-60
3. Fertilization	30-40
4. Watering	40-50
5. Collection of fallen materials	40-60
Haryana, 2011	

Gender Analysis and work participation: The term 'gender relation' refers to the relation of power between women and men as revealed in a range of practices, ideas and representation. These include division of labour, roles and resources between men and women. They also include ascribing to women and men in different abilities, attitudes, desires, personality trait and behavioural pattern. Gender analysis is also influenced by other social factors such as caste, class and race. Thus the gender analysis is not only subjected to biological factors but is result of social structures and processes. The gender differentiation emphasizes (i) sharing work (ii) sharing resource and benefit (iii) human rights and (iv) culture and religion. The participation of women in agricultural activities depends on situation. The activity profile of women in rural areas is divided into three types. These are, (i) Productive/Economic role (ii) Reproductive /family role and (iii) Community role.

1. **Productive/Economic Role:** These roles include tasks as, farming and trading that have direct economic benefits.
2. **Reproductive/family role:** These are also referred as household production role which includes
   (a) House building repair
   (b) Food preparation
   (c) Fuel collection (now in tribal locations)

(d) Water collection

(e) Child care and rearing

(f) Family health care

(g) House cleaning and sanitation.

3. **Community role:** This is the social role which includes organizing social and traditional functions and ceremonies and participating in village projects. It also includes political role such as becoming member of Panchayats. The role analysis of women indicates maximum involvement in household activities followed by farming activities.

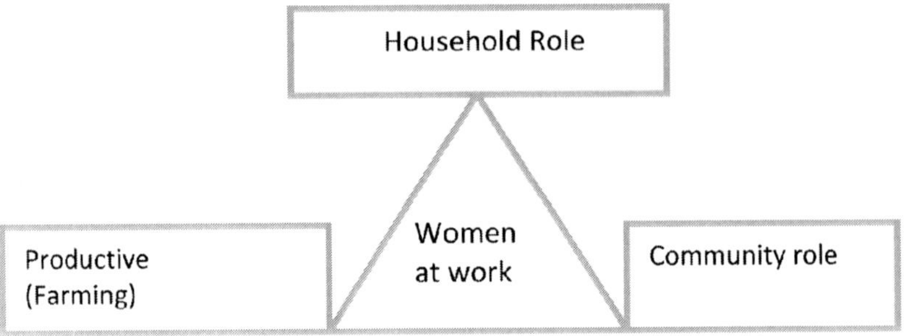

Participation of women in crop improvement Program: The participation of women in agricultural activities has been assessed covering five districts, namely Nayagarh, Puri, Cuttack, Dhenkanal and Angul with a randomized sample of 280 women which reveal the following information.

The participation of women in farming operation varies depending on location, crop, status of family, caste structure and many other socio-cultural factors. The women belonging to high status usually work inside family boundary mostly in the post harvest operation. But all the crops are not cultivated in the area and thereby the participation is subjected to growing of particular crop by the sample under study. The cropping intensity in Odisha is 158% (2010-11). The participation is based on the crops grown in the area. Vegetable is grown in the state with area of 133.12 thousand hectares with production of 9,261 TMT for the corresponding year. The cultivation of coriander, ginger and turmeric is confined to tribal districts. The women in tribal zone make full participation. The participation has been calculated taking frequency out of the total of 280 who constituted the sample.

The women in general do not like to disclose the information at the beginning even though they participate in all kinds of home and farm activities. Their participation may be categorized on the basis of work. The women agricultural laborers when step into the wage market they do have little choice in work participation. But in general there is social norm that in which type of works the women should work. The activities like ploughing, laddering, lift irrigation by country methods are the prohibited area for women because of use of physical strength. The work participation of surveyed sample is presented here but actually it is very difficult to quantify the participation of women in farming and related dimensions.

Table 2.2: Participation of women in crop improvement

S. No.	Crop	Mentions (f)	Percentage (%)
1.Cereals	1. Paddy	275	98.21
	2.. Millet	250	89.28
	3. Maize	158	56.42
	4. Ragi	175	62.50
2. Pulses	1. Green gram	140	50.00
	2. Black gram	135	48.21
	3. Red gram	110	39.28
	4. Horse gram	84	30.00
3. Oilseeds	1. Ground nut	120	42.85
	2. Mustard	95	33.92
	3. Sesamum	86	30.71
	4. Niger	90	32.14
	5. Sunflower	45	15.35

Contd.

S. No.	Crop	Mentions (f)	Percentage (%)
	6. Safflower	28	10.00
4. Commercial crops	1. Sugarcane	22	7.85
	2. Chili	65	23.21
	3. Onion	74	26.42
	4. Garlic	67	23.92
5. Vegetable	1. Cabbage	142	50.71
	2. Cauliflower	132	49.28
	3. Ladies finger	200	71.42
	4. Brinjal	192	68.57
	5. Tomato	135	48.21
	6. Carrot	64	22.85
	7. Pumpkin	250	89.28
	8. Beans	38	13.57
	9. Peas	12	4.28
	10. Raddish	132	47.14

In analyzing the work participation of the sample as per group of crops the following picture was observed.

Table 2.3: Comparative analysis of work participation of women

	Crop Group	Average percentage of Participation
1.	Cereals	60.97
2.	Pulses	41.87
3.	Oilseeds	27.40
4.	Commercial crops	20.35
5.	Vegetables	46.53

The results high light that women in farming participate more in cereal crops which constitute main food grain followed by vegetables which fetches money readily from market, pulses that families irrespective of size of holding preserve for family consumption. The oilseeds are cash crops and spices are location specific in which women participate is need based.

(i) **Reaction about present farming:** In a survey of Nayagarh district, the opinion of farm women in general irrespective of small, marginal or large farmer was analyzed. The actively involved women in farming feel that although productivity of certain crop has increased, the net profit is less and the soil and water are becoming polluted. Particularly the continuous application of fertilizer made soil infertile as a result the cost of production is increasing every year without proportionate increase in productivity. The overall reaction of 240 farm women spread of four blocks reveal interesting information.

Table 2.4: Reaction of farm women about present farming

	Farming at present	Agree	Don't agree	Total
1.	Profitable	60	180	240
2.	High productive	110	130	240
3.	Good family engagement	70	170	240
4.	Soil in good condition	50	190	240
5.	Produce is of good quality	75	165	240
6.	Cost of production is within affordability	45	195	240
7.	Less use of fertilizers and pesticides	40	200	240
8.	Good market value	80	160	240

Box No. 2.2. Women Perceived value about present farming

	Positive	Negative
1. Landless	20.00	80.00
2. Marginal women	35.00	65.00
3. Small	40.00	60.00
4. Large	50.00	50.00

The women view all the above attributes in negative way inferring that earlier farming seemed to be better than present. Again classifying women from small, marginal and large land holders, it seems that resource poor women have much negative value describing that at present farming is far away in fulfilling the requirements of rural households.

Taking all categories of women into consideration, the overall reaction is negative while it is more observed in case of landless, marginal and small compared to women of large farmers. It is unfortunate that inspite of so much efforts, the women express negative reaction about present farming system in which they work.

II Factors of Sustainability

The agriculture in general is viewed as non profitable for the reason that every year farmer faces natural calamity and lack of good market for the produce. It is also expressed that with increasing cost of seeds, fertilizers, pesticides and labor, the small and marginal farmers feel better to earn wages rather devoting money, time and labor in agriculture which do not compensate their expenditure. Many of women also feel that productivity of land has been lost as it does not respond to locally available compost and only to chemical fertilizers the cost of which is increasing continuously.

The reaction of sample women with regard to factors of sustainable farming for good production and productivity, the ranking of factors was obtained as follows.

Table 2.5: Ranking of sustainable factors of agriculture

	Factors	Mean Score	Rank
1.	Ecologically sound	4.56	III
2.	Economically viable	5.87	I
3.	Socially demanded	3.62	IV
4.	Locally adaptable	4.96	II

(i) Economic viability: Price for organic food is same as conventional foods in terms of growing, harvesting, transportation and storage. Studies have shown that organic agriculture is commercially viable, that farmers can achieve more income because of premium that they need fewer inputs to maintain return. Organic system are based on the optimum use of local resources and technologies that can give farmers greater independence and more control over their means of production. The farmers enter organic farming to make holistic approach. Other considerations include environmental and social concerns, economic necessity, lack of chemical inputs and market demands. Small farmers are encouraged to take up organic farming to produce more and get export price. To the small, marginal and big farmers sustainable agriculture means sufficient production and minimum risk along with conserving resources. Women feel that it is the economic factor which makes agriculture sustainable. The present income generation from land is much less compared to other walks of living. With expansion of road and building work in rural areas, laborers are earning much more with less physical labor than the farmers and farm women. Unless earning from land may be of any crop, fruit or vegetable is able to sustain the family why they would remain in farming. If economic points are taken care of, then the agriculture would be sustainable and rural families would continue to work in farm. The cost of production, money value in the market and net income received by the family is very marginal which does not support the family to live.

Economic viability is measured by the variables like production level, conservation of natural resources, profitability, adequate income and cost of technology. The recorded reactions of the sample with regard to components of economic viability in their assessment are contained in following table.

Table 2.6: Reactions about components of economic viability

Factors	Fully satisfied		Satisfied		Not satisfied	
	f	%	f	%	f	%
1. Production level	50	20.83	74	30.83	116	48.34
2. Conservation of natural resources	21	8.75	45	18.75	174	72.50
3. Profitability	28	11.67	58	24.16	154	64.17
4. Adequacy of income	20	8.33	48	20.33	172	71.34
5. Affordability for technology to adopt	55	22.91	61	25.42	124	51.67

Satisfaction of sample women with regard to economic viability of present farming is far from expectation. Productivity level of cereals has increased while in case of pulses, oil seeds and millets it is not very much visible. In growing of vegetables, hybrid seeds of vegetables, like brinjal, tomato, cauliflower, cabbage, ladies finger have shown very good performance in terms of productivity but at the same time incidence of disease and pest, lack of storage facilities have not helped farmers to gain much. Example of throw away sales of vegetables in the state is very common. Moreover, market structure is such that middle men get more profit than the producers. The data are indicative of facts that present agriculture is not very much attractive to keep farmers in farming, rather farmer bear loss in many places for produce. The non-payment of loan taken for agriculture is accumulating in very big way in rural areas.

(ii) Locally adaptability: Locally adoptability is the degree to which technologies are introduced and integrated into existing farming system of the households. There are many farm technologies which are quite suitable for agro-climatic condition. At micro-level when individual farm woman considers for adoption, she looks forward to the factors like (i) relative advantages in terms of production and labor involvement (ii) compatibility to the existing farm condition (iii) complexity in adoption owing to availability and use of required inputs (iv) affordability and (v) local market demand. In the opinion of sample women, the innovation or new technologies can be sustainable provided these factors are satisfied.

Sample women consisting of landless, marginal and small farm holders view that sustainable agriculture can be achieved when new technology integrates with existing farming system, i.e. kharif and rabi season adjusting to time of seed bed raising or sowing as the case may be. Market demand and production at household level can decide the sustainability of agriculture in particular geographical areas. With constant change in the preference of consumer which is market linked also act as determinant for sustainability of technology. Disaster like heavy untimely rain, unexpected flood, high temperature which are frequent now also add to continuation of agricultural sustainability. Many sample women reported about disasters that they face almost every year. Dry spell and late rain often make farming difficult. During Kharif season many are forced to keep land fallow.

Table 2.7: Components of adaptability of sustainable agriculture practices

	Components of adaptability	Mean Score	Rank
1.	New Technology	4.85	I
2.	Government Policy	3.43	IV
3.	Consumer choice	4.15	III
4.	Market demand	4.39	II
5.	Disaster mitigation	3.27	V

New technology, market demand, consumer choice, govt. policy and disaster mitigation are the crucial factors that decide the adoption of practices of sustainable agriculture.

(iii) Ecologically sound: The sustainable agriculture should be ecologically sound. We are faced with three major issues, (i) environmental degradation (ii) more food for increasing population and (iii) profit at farmer level. Green technologies as an integrated farming system considered to have promising traits to meet these requirements. The gravity of the environmental degradation, arising from faulty practices have drawn the attention for ecological sound, viable and sustainable farming system. The shift from chemical to ecological agriculture should be gradual. A sudden switch over may lead to disaster and discourage the farmers towards sustainable farming. Ecological farming can be sustainable and profitable. It is an integrated system and knowledge intensive practice. It is commercially attractive to growers compared to inorganic farming. It has been observed that chemical farming, fertilizer and plant protection account for nearly 30% of the total cost per hectare. Bio-fertilizers and organic content can very well reduce the cost of cultivation.

Eco farming is based on the considerations like, crop, soil, climate of the region which should be in one line of approach. Eco farming implies that farming regions and individual farms must be treated as ecological system; Ecology is the science which deals with the relationship between the organisms and their environment. But the environment is not only confined to the natural conditions, i.e. soil and climate. It encompasses entire complex of physical, economic, social and cultural conditions which affect the growth and development of organic system.

Objectives of Ecological farming: According to Sankaram (1996) this technology may be grouped into four categories.

1. Those that reduce environment burden of green house gases (CO and CH), global warming and ozone depletion. This needs attention to promote renewable energy like, draught animal power, electricity, garbage disposal, biogas from organic wastes.
2. Those that reduce demand on land, water and biodiversity without adverse effect on agriculture production and nutritive value of food. It requires soil health, change in cropping pattern to maximize ecological productive efficiency, improve water use efficiency through conjunctive use of rain water, tank, under ground, well and river water and reduces conveyance losses.
3. Those that continue to improve crop productivity, under shrinking land resources, hybrid vigour, gene pyramiding, multiple cropping patterns, integrated nutrient management and integrated pest management.
4. Reduce hunger and poverty, adopt, cost effective farming to bring equity of food price and wage, encourage job, promoted growth right in the village to arrest migration as ecological refugees.

Women feel that present farming practices are not better than the past practices. The quality of natural resources maintained is not good. The change in climate particularly frequent erratic rain, high temperature and sudden rainfall make all farms planning futile. They feel that change in climate specifically rain fall changes sowing time, timely cultural operation and harvesting time putting family in great difficulty. All these deviations lead to loss in income. Even not directly, the expression of women certainly hints at ill effect of ecology in the area. Ecological impact is very much realized by all categories of farmers. Climate change and irregularities in farming operation is now well experienced by the farmers. The reaction of women sample with regard to various components of ecological issues is recorded in the box.

Box No. 2.3: Ecological elements

		Reaction mix
1. Soil	-	Poor, needs more and more fertilizer every year. Without fertilizer there is hardly any production. The soil is becoming stiff for which good and pulverized field is not possible. In irrigated areas due to intense use of tractors, the soil is not settling down for transplanting of paddy. In summer big cracks are noticed.
2. Water	-	Water availability is most uncertain. In rain-fed areas, water hardly remains within reach zone of roots. Dry spell, spoilage of seedlings is very common. In the same time water logging condition in low land does not permit sowing and application of fertilizers in time. Water holding capacity of soil has gone down. Excess water in irrigated areas becomes problem. Field remains moist and muddy. Too much irrigation often does not permit to takeup pulse crops.
3. Soil nutrients	-	Compost and green manure are hardly supplied to soil. Major dependence remains on chemical fertilizer that too only NPK and never micronutrients. One has to apply more fertilizer to increase productivity level. Soil testing is hardly practiced in survey areas.
4. Crop rotation	-	Crop rotation is very limited. In canal irrigated area only paddy after paddy is cultivated, in rain-fed areas second crop is taken if there is soil moisture otherwise land remains fallow. Rainfall at end of October helps crop rotation and taking up of second crop.
5. Use of organic manure	-	Organic manure is very limited. Small and marginal farmers use organic manure in paddy and to some extent in vegetable. Application of it per acre is very less. One can hardly find compost pit in the back yard of farm families as per recommendation.
6. Use of bio-pesticides	-	Bio-pesticides have not yet created confidence in the minds of the farm families. It has limited use and more in vegetables while paddy, ragi and other cereals go with chemical pesticides.

Socially demanded: The innovation or change in agriculture should empower all households for equal accessibility to community properties. Techniques of land use, adequate capital and technical assistance when made available to all households, the farming system remains sustainable. The social variables in rural areas have differential impact owing to caste, religion, local habits and other factors. The accessibility of women to different common resources of the community is regulated by certain norms fixed by the village heads. During data collection it was observed that in certain locations, women were afraid of expressing their views rather need the approval of the village leaders. However, the views of sample in respect to sub-components of social factor are quite interesting and educative.

Table 2.8: Reaction about social factors

	Sub-Components	Much more	More	Little
1.	Accessibility to community resources		√	
2.	Use of energy in agriculture		√	
3.	Technical support to all		√	
4.	Availability of capital		√	
5.	Marketing opportunity			√
6.	Participation in decision making		√	

In social parameters except opportunity for good marketing, the women are in favor of existing farming system. The present agriculture appears to be quite positive to sustainability. Availability of capital in farming although is not a problem but arrangement of guarantee for availing it is problem at many places stand as barrier. The time has changed and rigidity of social system is now accommodative. All the villagers are equally enjoying community properties with little deviations here and there. Use of energy in agriculture is gaining momentum in the state. The use of tractor, power tiller, harvester, water pump, bore well *etc.* is in increasing trend. Use of electricity, diesel, petrol *etc* in farm sector is increasing. Extension support to all farmers and farm women is extended and farm technology is available not only through extension personnel but also through different media. Likewise farm loan is now easily available in rural areas. Women are now taking active part in farm decision making process. The social parameters and sustainable agriculture are closely associated. However, the sample with following characteristics was found to be more associated with farm sustainability. The social variables are in the process of change like economic ones.

Box No.2.4. Social variables

	'r' value
1. Education	0.854
2. Training	0.642
3. Social participation	0.348
4. Membership in formal organization	0.410
5. Contact with extension agencies	0.470

The social just or variable for sustainable agriculture is quite favorable in the area of survey. Education of women, training undergone, social participation, membership in formal organization and contact with extension agencies are significantly associated with feeling of rural women looking forward for sustainable agriculture.

III Gap in Awareness about Advantages of Sustainable Agriculture

The rural women are acquainted with traditional agriculture which has much relevance with sustainable agriculture. But the scientific part of it and real benefits are not known to many. The sample women in a group discussion about advantages of sustainable agriculture expressed their know- how as has been recorded.

Table 2.9: Expression about benefits of sustainable agriculture

	Advantages	Know	Don't know	Mean score	Gap (%)
1.	Diversity of crop and produce	120	120	0.50	50.00
2.	Maintains soil fertility	60	180	0.250	75.00
3.	Improves physical, chemical and bio-logical features of soil	45	195	0.187	81.30
4.	Makes use of local resources	136	104	0.	44.00
5.	Reduces use of external inputs	128	112	0.270	47.00
6.	Recycles resource for better use	80	160	0.333	66.70
7.	Reserves biodiversity	61	179	0.254	74.60
8.	Crops tolerance to stress condition	30	210	0.125	87.50
9.	Helps in pest and weed control	58	182	0.241	85.90
10.	Makes use of ITK	40	200	0.166	83.40

Advantages of sustainable agriculture are many and women know them in general. But the scientific reasons and visible impact are not known to many. In calculating the gap in ideas about the benefits of sustainable agriculture reveals that gap to a varying degree. More gap is observed in case of tolerance to stress conditions, control of pest and diseases, use of ITKs, improvement of soil composition and maintaining of soil fertility.

(i) Tolerance to stress condition: Stress condition occurs when rain fall is delayed or most erratic. The problems faced in germination dry spell, burning of seedlings and stunted growth of crop plants are some of such situations. Sustainable Agriculture could be used to solve problems are not convincible to farm women. On a field observation taking stress conditions of the crops and comparing application of compost and inorganic manures the results were found to be convincing.

Box 2.5. Growth parameters		
	With compost	**With inorganic fertilizers**
1. Germination	More	Less
2. Crop condition	Good	Stunted
3. Weed	Less	More
4. Leaf burning	Much less	Much more

The sample women on a field visit and group discussion (Nayagarh) lastly agreed to the good effect of compost over inorganic fertilizer in field condition with reference to germination, crop rotation, weed population and leaf burning. Our farmers and farm women are so much keen to inorganic farming that they do not easily agree with visual effect of organic farming.

Box No.2.6. Consumption of pesticides per hectare	
Year	**Odisha (gms.ha.)**
2004-05	118.00
2005-06	138.53
2006-07	148.94
2007-08	148.24

Source: (Odisha Agril. Statistics, 2009-10)

(ii) Pest and weed control: Sustainable Agriculture controls weeds and pests to a considerable extent. The women do not find much difference between present farming system and sustainable agriculture in controlling pest. Weed control in paddy, ground nut and other crops is a problem in the state. The use of pesticides and weedcides is in increasing trend. However, specific experiment is required to show that in case of sustainable agriculture whether the weed infestation and pest attack remains under the control. The use of ITK is found to be effective in controlling weeds such as application of mango leave slurry and keeping water in crop field for a period of three days.

(iii) Make use of ITK: The women are practicing ITKs in number of crops. These are ash dust, mustard oil in store grain, cow dung slurry, *neem* cakes etc. But these ITK_s have not been standardized in terms of dose, method and interval of use. Once the sustainable agriculture is being adopted, the ITKs not used or forgotten will come to practice. The study indicated that higher the formal education of women greater the use of chemicals. The chemical inputs are readily available and easy to use.

Box No. 2.7.Principal crops

	% share
1. Rice	46.00
2. Other cereals	5.00
3. Pulses	23.00
4. Oilseeds	8.00
5. Fibers	1.00
6. Sugar cane	1.00
7. Vegetables	8.00
8. Spices	2.00
9. Others	6.00
Total	100.00

Sources: Orissa Agril. Statistics 2010-11

(iv) Diversity of crops and product: Sustainable agriculture goes with rotation of crop and growing of different crops. In this case the farm family decides two important issues, i.e. basic needs of family (Rice) and market demanded product. During survey we observed that it was women who could reveal the family requirements and marketable produce more than men. State produces a variety of farm produces in different locations as per agro-climatic zones. The major crops of the state are, cereals (paddy, millet), pulses (green gram, black gram, arhar) oilseeds (mustard, ground nut, sesame, castor, Niger, sunflower and safflowers) Among oilseed crops , Niger has special place in tribal areas. The commercial crops, like soybean, sugarcane, potato, chilly, onion etc. are grown to a significant level. The participation of women in all these crop production has been well recognized. Our agriculture is equally shared by men and women while at post harvest, it is the women who play major role than counterparts. Diversity in crop, variety and products ensure against total failure and better market value.

(v) Improved soil structure: Sustainable farming improves physical, biological and chemical structure of soil. It is common habit with rural women to put all family waste and waste of cattle in the field. Many households dig a pit in back yard and put all garbage in pit till it is filled and take to field before sowing of seeds. Thus in sustainable agriculture the women are actively involved. Compost, vermi-compost, azoles etc. have added strength to our sustainable agriculture. Around each KVK of the country, vermi- compost has been taken up in large scale adding to organic farming.

Box No. 2.8. Textural classes of soil in Odisha

Textural class	% Sand	% Silt	%Clay
1. Clay	20	20	60
2. Silt loam	20	70	10
3. Sandy loam	65	25	10
4. Loam	40	40	20

(vi) Reserve biodiversity: Biodiversity is the present concern of agricultural scientists. Due to change in climate many crops are not in practices. Most biologists accept the estimate of Edward O. Wilson that earth is loosing approximately 27,000 species per year. The estimate is based on disaapearance of ecosystem in tropical forest and grass land.The single greatest threat to global biodiversity is the human destruction of natural habitats. Preservation of biodiversity is the responsibility of all the nations in the world. In the last three decades, focus has shifted from preservation of individual species to the protection of large tracts of habitats linked by corridors that enable animals to move between habitats. Presevation of biodiveversity at the molecular level is preservation of genetic ones. All over world, attempt is made to collect and preserve endangered organisms. Today, 40,000 species of plants and animals are dauily used.

Box No.2.9. Plant species spectrum	
Species	**Number**
1. Existence	240,000
2. Endangering species	20-25,000
3. Analyzed species	5,000
4. Cultivable species	3,000
5. Caloric species	30
6. Consumption species	4

(*Source:* T.N.Khoshoo 1992)

The crop coverage of the state of Odisha depends on soil, climate, local requirements and many other factors. The area under different crops of the state from 2005 to 2010 is furnished herewith

Table 2.10: Crop coverage in Odisha (ha)

Crops	2005	2006	2007	2008	2009	2010
Maize	15962	17627	15939	16225	17025	22712
Ragi	731	561	2267	2272	2606	2540
Mung	10778	8715	9003	22694	22996	8980
Biri	6828	7433	16504	-	7582	6781
Arhar	13689	12875	13494	14520	15214	17664
Cowpea	403	6715	20584	1628	1931	2256
Groundnut	8395	6925	11292	11548	12106	8950
Til	6068	1823	11693	12648	12924	9640
Sunflower	6	0	0	8	0	0
Fiber	14282	6687	3036	3971	3085	23485
Sugarcane	12792	4244	5224	5854	5871	5836
Total Vegetable	19109	15005	25329	34547	30987	32520
Total Spices	3124	2504	2603	2928	3095	3125
Other Horticulture crops	4518	2816	-	2578	4107	16880

(*Source:* Agril. Statistics, Odisha 2011-12)

The area under different crops covered in Odisha from 2005 to 2010 reveals the change in coverage due climatic effect, market demand and preference of consumers. Due to less cold in Odisha the wheat is almost out of cultivation.

The crop substitution in Odisha from 2005 to 2010 reveals the increse and decrease of area under different crops.

Box No.2.10. Biodiversity of germ plasm collection in agriculture

Crop	Number
Rice	40,000
Wheat	40,650
Pulses	41,350
Oilseeds	30,415
Cotton	20,755
Coffee	460
Tea	1,725
Vegetables (Summer)	3,650
Vegetables (Winter)	1,850
Total	2,16,075

(*Source:* Desai,B.K.& Pujari,B.T. 2007,Sustainable agriculture: A vision for future NIPA,p.78)

Use of local resources: Agricultural crops need nutrients containing N, P, K and minor elements. All these are now supplemented by inorganic fertilizer. The bio-fertilizers are now gaining popularity among farmers. The women farmers are also involved in preparation and use of bio-fertilizers. The local resources are straw, grasses, kitchen waste, village wastes etc. LEISA (Low External Input Supply Agriculture) is a concept that highlights use of local resources. LEISA seeks to optimize the use of locally available resources by maximizing the complementary and supplementary role of locally available resources.

Agro-wastes are the local resources which can add organic matter to the soil. In rural areas, agro-wastes are not properly used. Once the women are aware and trained as how to make use of agro-wastes, we can add huge quantities of organic matter to the soil.

Synergistic effects of different components of the farming system. External inputs act as complementary components. The local resources like poultry litters, legumes, compost and subterranean clover can provide Nitrogen to soil and crops.

LEISA: (Low External Input Support Agriculture)

Box No. 2.11. The ecological, economic and social criteria of LEISA

Ecological Criteria	Economic Criteria	Economic riteria
1. Balance use of nutrients and organic matter	1. Sustained farmer livelihood system	1. Wide spread equitable adoption potential
2. Efficient use of water resource	2. Competitiveness	2. Reduced dependencyon external input
3. Diversity of genetic resources	3. Efficient use of production factors	3. Enhanced food security
4. Efficient of genetic resources	4. Low relative value of external inputs	4. Respect to indigenous knowledge
5. Efficient use of energy sources		5. Employment generation
6. Minimal negative environmental effect		
7. Minimal use of external inputs		

Low External Input Supply Agriculture (LEISA):- The term Low input agriculture has been defined as a production activity that uses synthetic fertilizers or pesticides below rate commonly recommended by Extension Services. It does mean elimination of these materials. Increase in yield is maintained through greater emphasis on cultural practices, IPM and utilization of on-farm resources with better management process. The LEISA concept stresses on optimum use of locally available resources and use of external input as complementary to the system.

Agro wastes are plenty in villages. To make our agriculture sustainable, there is need to pay attention for proper use of agro-wastes. The C.N.ratio of different agro wastes is given.

Table 2.11: Available Agro-wastes and their C. N. ratio

Si. No.	Agro wastes	C. N. Ratio
1.	Rice straw	80:1
2.	Wheat straw	80:1
3.	Barley straw	100:1
4.	Maize stalk and leaves	50:1
5.	Cotton stalk	70:1
6.	Sugar cane trash	120:1
7.	Lucerne	19:1
8.	Soybean residue	45:1
9.	Groundnut foliage	20:1
10.	Jute leaves	25:1
11.	Green weeds	13:1
12.	Grass clippings	19:1
13.	Water hyacinth	29:1
14.	Ipomoea	43:1
15.	Coir pith	112:1
16.	News paper	100-800:1
17.	Card board	400-563:1
18.	Vegetable trimmings	25:1
19.	Food scraps	15:1
20.	Leaves and foliage	60:1

(*Source:* Barik,T. Gulati, J.L.M. Production of vermin compost from agricultural waste, Recent development in organic farming OUAT, p.217)

Cited agro-wastes are available in rural areas but their proper use has not been done yet to meet the organic needs of crops. Women although know these wastes still they are not trained to take advantages of wastes in agriculture. Rice straw, maize stalk, groundnut foliage, green gram and black gram foliage, green weeds, water hyacinth, ipomeas, coir pith, leaves and foliage are not properly converted to required quality of compost.

In rural areas each and every farm families used to have compost pits in back yard where they put all garbage to make compost. The compost made out of these wastes is taken to field before summer cultivation in rice growing areas. With change in time and increased dependency on inorganic fertilizers, the compost making has been neglected. The scenario of compost making of the sample is observed to be as follows.

Table 2.12: Compost making sample scenario by women N=280

Si. No.	Particulars	Number	Percentage
1	Have regular compost pits	250	89.28
2.	Compost pit made as per recommendation	26	9.28
3.	Regular application of compost in field	165	58.92
4.	Training received on compost making	28	10.00
5.	Regular habit of putting garbage in compost pits	32	11.42

Compost is a must for sustainable agriculture. Our farmers have to go long in replacing inorganic fertilizers by organic approach.

Agro-Organic wastes in India:

There are various estimates of agricultural wastes availability. These are based on crop residues, net availability on remote sensing techniques. It is suggested that average value for crop wastes is 350mt and that of animal it is 650mt.Hence around 1000mt of agricultural wastes are available in the country.

Box No. 2.12 Agricultural wastes	
Quantity (mt)	
1. Crop wastes	301.6
2. Animal wastes	944.4
3. Crop residue including industrial wastes	108.0
4. Crop wastes cereals only	289.7
5. Crop residues	407.0
6. Animal wastes	2018.0
7. Crop residue including industrial wastes	347.2
8. Crop waste residues	287.3
9. Animal wastes	633.0

(*Source:* Sharma, A.K.2008 A Handbook of Organic Farming, Agrobios . 146)

In rural areas there are huge agricultural wastes which can be profitably converted to good compost. Our approach has not been very much effective in this direction. On an interaction with extension personnel like VAW and Agriculture Extension

Officers, it was ascertained that now- a -days their focus is to popularize high yielding or hybrid varieties of crop with recommended input use which are mostly inorganic in nature.

The rural women are well acquainted with agro-wastes. They are to be motivated through effective extension approach to prepare good compost in their back yards. The methods the women prefer for motivation are found to be (i) demonstration (ii) group discussion (iii) individual contact and (iv) visit of compost pits in nearest locality.

Organic Farming and Food Security: FAO has cautioned that though organic farming helps produce nutritious food and represent a growing source of income for developed and developing countries, it alone cannot ensure global food security. As per its report roughly 2% of the world's crop land was farmed organically in 2005 and in 2006, organic produce generated US 24 billion dollars in sales in the European Unions, the US, Canada and Asia. The argument is that organic farming cannot substitute conventional farming to feed world's growing population. We can only take up organic agriculture in targeted crop in targeted areas beside continuing use of chemical inputs.

Organic farming avoids use of synthetically fertilizers and pesticides. It heavily depends on crop rotation, use of crop residues, animal manures, legumes, green manure, off farm organic wastes, bio-fertilizers, mechanical cultivation and biological control of insects.

The answers to the question as why we should opt for organic farming, the response becomes positive in the hypothesis that organic farming can increase farm productivity, repair decades of environmental degradation and knit small farm families into more sustainable distribution leading to improved food security. Organic farming is one way to promote either self-sufficiency or food security.

The holistic production management system which is called sustainable agriculture promotes and enhances, agro-ecosystem health, including bio-diversity, biological cycle and soil biological activity. Many experiments reveal that organic farming can produce more than conventional farming. The enlisted benefits for which we advocate organic farming are,

1. Maintains environmental health and reduces pollution.
2. Reduces human and animal health hazards
3. Keeps agricultural production at sustainable level.
4. Reduces cost of crop production
5. Ensures optimum utilization of natural resources
6. Reduces risk of crop failure
7. Improves soil physical properties
8. Improves soil chemical properties

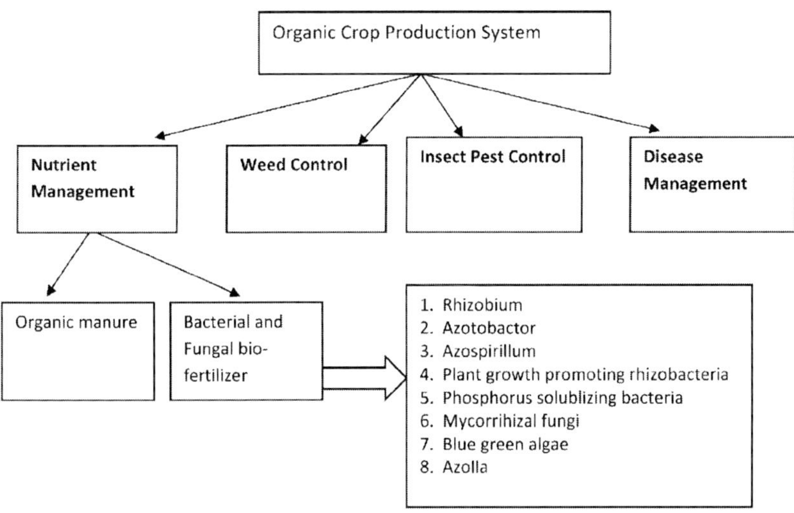

Fig. 2.1: Organic Crop Production System

Food for all and Role of Women farmers: According to the report of the United Nations Secretary General (2011), of the 900 million people who are likely to be in extreme poverty in 2015, India is expected to be home to more than 300 millions. Our biggest challenge still remains ensuring food and nutritional security to our masses. It is the right of every person to have regular access to sufficient, nutritionally adequate and culturally acceptable food for an active and healthy life. According to International food policy Research Institute's Global Hunger Index of 2010, India is ranked a poor 67th in battling hunger. It is among 29 countries with the highest level of hunger. In India, availability of food remains the focus to the family as food consumes the bulk of the family income in majority of houses. Nationwide, 57% of the expenditure in rural areas and 44.4% in urban areas is used for the purchase of food.

Food security is a complex issue involving number of dimensions. In simple words; food security is defined as economic access to food along with food production and food availability. FAO has enlisted three specific aims in the concept of food security.

1. Ensuring production of adequate food supplies.
2. Maximizing stability in the flow of supplies.
3. Securing access to available supplies on the part of those who need them.

Dr. Swaminathan includes three components in the term of food security. These are (i) food availability (ii) food accessibility and (iii) food absorption which depends on safe drinking water, environmental hygiene, primary health care and education. An important issue related to food security is that poor persons should have purchasing power to have access to sufficient and nutritious food grains. Mere availability of food grain does not ensure food security.

Challenges of food security: It is a decided fact that there is need for multi pronged strategies for food security which include,

1. Sustainable management of natural resource
2. Removal of trade barriers
3. Equitable access to land adoption of sustainable technology
4. Social sector investment on health, education and infrastructure
5. Improved governance

Table 2.13: Food grain-consumption gap in major food items in India (million tons)

Year	Rice	Wheat	Pulses	Sugar	Oilseeds
2005-06	2.3	-4.2	-0.9	-4.1	-4.6
2006-07	6.7	-o.4	-1.0	0.7	-1.9
2007-08	6.6	2.4	-1.6	8.3	-6.2
2008-09	6.2	2.2	-2.6	5.1	-4.0
2009-10	6.0	9.8	-2.2	-8.1	-4.7
2010-11	-2.8	2.1	-2.3	-7.5	-6.0

(*Source:* RBI Bulletin 2010)

Need for food security should be considered not only from quantitative aspect but also from qualitative aspect. There are some qualitative indicators justifying priority of food security at national level.

1. According to Global Hunger Index, 2009-10 India ranks 96 in a group of developing countries.
2. According to World Food Program, nearly 50% of the world's hungry live in India.
3. About 35% of our population over 350 million is food insecure, consuming less than 80% of the minimum energy requirements.

4. Nearly 9 out of 10 pregnant women between 15-40 years are malnourished and anaemic.
5. Anemia in pregnant women causes 20% infant mortality
6. India contributes 21.6% of total death in the world below 5 years age group.
7. Malnutrition accounts for 50% of under than 5 years old death.
8. About one third of under weight children under five live in India. 9.51 % of children are not immunized.

CHAPTER 3

Adoption of Sustainable Practices in Agriculture

Sustainable agriculture requires sincere attempt by the farmers to be sustainable in their approach. The practices are neither new nor difficult. But the minds of the farmers are so much inclined to inorganic farming that it is not so easy to divert their minds for sustainable agriculture.

The important Practices of Sustainable Agriculture are

I. Mixed Farming

It is the combination of two independent agricultural enterprises on the same farm. It may be crop with livestock and can be termed as diversified farm. Mixed farming is to support or add to farm income. Further, in mixed farming contribution of livestock varies from 10 to 49%. When livestock is cow or buffalo and contribution within the cited range, it is called mixed farming and when livestock is more than cow and buffalo may be poultry it is termed as diversified farm.

Table 3.1: Advantages and disadvantages of mixed farming

Advantages	Disadvantages
1. Highest return from farm	1. Indigenous method till now
2. Provides work for entire year	2. Animals are to be sold when not productive
3. Efficient utilization of land, labor, equipments	3. Needs healthy for replacement
4. Crop byproduct used by livestock which in turn gives milk	
5. Manures available for soil	
6. Supplies needed food	
7. Scope for intensive farming	
8. Provides alternative for income	
9. Provides draft animals	
10. An indicator of social status	

Requirements of Mixed Farming

1. Management Practices
2. Sound Cropping Scheme
3. Good livestock
4. Transport facility
5. Marketing facility

Women and Mixed Farming: Women in rural areas are intimately associated with farming system. Each and every rural farm family possesses domestic animals to add to their family incomes. Even they do not directly respond to mixed farming. The reality is that they keep domestic animals and supplement to family income and manure to the crop field. The normal mixed farm and diversified farms are observed in the following combinations.

1. Agriculture + Animal Husbandry
2. Agriculture + Animal Husbandry + Fisheries
3. Agriculture + Animal Husbandry + Fisheries + Off farm occupation

Most common livelihood system of small and marginal farmers is crop and animal husbandry. The knowledge of women respondents (N=150) about mixed farming and its benefits are found as follows:

Box No 3.1 Mixed farming

Particulars	Percentage
1. Know the benefits	30.00
2. Interest to have mixed farming	45.00
3. Adopting mixed farming	24.00
4. Intends for mixed farming	12.00

Mixed farming is practiced in rural areas. Now the farmers are keeping jersey cows and other breeds for milk purpose because there exists a good market for milk. In addition to crop, milk production constitutes a good percentage of family income. The other animals like goat, sheep and poultry bird is common in the villages. Although these elements constitute mixed farming, still the benefit of it is not fully realized by the farm families. Women take major role in animal husbandry. They undertake all kinds of animal husbandry activities like, feeding, management, collection of feed, growing of fodder etc. In real sense the participation of women in animal keeping is much higher than men.

Bullocks are the important means for cultivation of crops. Bullocks used to be main farm power for all kinds of farm operations mainly, ploughing, laddering, transport of manure, crop and other related operations. With introduction of tractor and power tillers, the bullocks are gradually replaced from farm operation. As per expression of farmers, maintenance of bullock throughout the year is more costly than hiring tractor or power tillers to meet the needs. Now farmers are not in favor of keeping bullocks. By the way we are losing cow dung manure which constitutes major part of our compost.

In an unit of mixed farming in case of small and marginal farmers, the income from milk is more and regular with less climatic hazards. People are interested to accept animal husbandry related technology by spending money because dairy ensures regular flow of income which is not the case for crop.

II. Crop Rotation

Crop rotation is the practice of growing a series of dissimilar/ different crops in the same area in sequential seasons. It confers various benefits to soil like replenishment of nitrogen through the use of green manure in sequence with cereals and other crops. Crop rotation also mitigates the build-up of pathogens and pests that often occur when one species is continuously cropped and improve soil structure and fertility because of deep rooted and shallow rooted crops. The nature of soil, climate and rainfall decide the types of crops to be crop rotation. It may include two or more crops. Benefits of crop rotation include (i) greater use of on-farm nutrients, (ii) mixing of crops slows down the spread of pests (iii) 10-25% increase in yield (iv) checking of soil erosion and (v) overall risk is distributed by different crops (vi) greater yield stability (vii) reduction in disease and pest incidence (viii) increase in arthropod diversity (ix) recycling of nutrients reserve from depth in soil and (x) transfer of N from N fixing species.

Box No. 3.2. Ingredients to avoid in composting
1. Residue sprayed with pesticides
2. Diseased with rust and viruses
3. Meat scrap
4. Hard prickle
5. Persistent perennial weeds

Further choice of crops and varieties is decided on (i) maintaining of soil fertility (ii) reducing of nitrate leaching (iii) reducing of weeds, pest and disease problem.

Important crop rotation

1. Black gram - Toria - Wheat
2. Pigeon pea - Wheat
3. Maize - Mustard - Urad bean
4. Rice - Urad bean
5. Maize - Safflower
6. Rice - Chickpea
7. Black gram - Mustard
8. Ground nut - Safflower
9. Rice - Mustard
10. Rice - Ground nut

Crop rotation is faced with certain constraints. In case of canal irrigation, the farmers are forced to grow paddy after paddy as other crops cannot be taken up. With availability of irrigation, farmers go for high value crop and that too field becomes unsuitable for pulses. In case of lift irrigation the crop rotation can be taken up systematically. In case of dry land area, monoculture is practiced as moisture is not available for subsequent crops. A large area of Odisha remains under monoculture that too only paddy.

Women in farming have expressed during study that due to change in rainfall pattern, crop rotation is faced with problems like at the time of harvesting of pulses in October. Rainfall makes difficulty in storing and harvesting of pulses mainly black gram in large of mid table central land zone in the state.

III. Compost

Compost is the heart of organic farming. Composting is a biological decomposition process that converts organic matter to stable humus like product under control condition. During composting process microorganism utilize decomposable microbial subtracts present in organic compost both as an energy source and for conversion to microbial substances. The importance of compost is multidimensional and very much required for any kind of farming that we practice. These are in short,

1. Finished compost has fertilizer value
2. It is an excellent soil conditioning agent
3. It improves texture, permeability and water holding capacity of soil
4. Increases water retention capacity and absorption capacity of soil
5. Improves fertility of marginal arable land for restoring of eroded land
6. Can be used as mulch in vegetable farming
7. Excellent materials for bedding

Compost Design: Compost design depends on type and amount of compost required, climate, availability of land, handling, labor and water availability.

Table 3.2: Materials used for composting

Sl. No.	Materials	Conversion ratio
1	Urine	2:1
2.	Dried blood	4:1
3.	Pig manure	5:1
4.	Poultry manure	10:1
5.	Comfrey	10:1
6.	Lawn trimmings	12:1
7.	Kitchen compost	12:1
8.	Farmyard manure	14:1
9.	Seaweed	19:1
10.	Garden compost	20:1
11.	Coffee grounds	20:1
12.	Horse manure	25:1

Contd.

S. No.	Materials	Conversion ratio
13.	Weeds	30:1
14.	Bracken	48:1
15.	Straw	80:1
16.	Woody pruning	100:1
17.	Bark	100:1
18.	News paper	200-500:1
19	Cardboard	-
20	Sawdust	500:1

IV. Green Manure

Green manures can be defined as any crop or plant grown and ploughed into soil to improve soil fertility by addition of organic matter and nitrogen. For green manure fast growing legumes with more vegetative growth should be used.

The choice of appropriate green manure depends on (i) suitability to climate (ii) fast growing and leafy green manure (iii) not closely related to following crop (iv) availability of seeds (v) length of time to grow and (vi) legumes are the best.

Green manure is obtained by two methods, (i) by growing green manure crops and (ii) collecting leaves from plants grown in wasteland.

Benefits of green manure are immense. These are (i) enhance soil fertility (ii) supplement for nutrition (iii) improve soil structure (iv) prevent soil erosion and (v) weed control .

(A) Green manure crops: Green manure crops are buried in soil by ploughings. These are:

1. Dhanicha *(Sesbania aculeta)*
2. Sesbanis speciousa
3. Sesbanis rostrata
4. Sunhemp (Crotalariajuncea)

5. Wild indigo (Tephrosia purpurea)
6. Pillipesara (Pheseolus trilobus)

(B) Green leave manure: Tender green twigs and leaves are incorporated in soil. These are:

1. Glyricida
2. Karanja
3. Neem
4. Calatrophis

For preparing compost, the farmers are advised to raise green leave manure plants on raised border of the crop and incorporate them before transplantation operation specifically in case of rice. The practice gained popularity among the farmers but due to use of inorganic fertilizers many farmers have discontinued the practice. It was abundantly observed in both sides of Hirakud command areas and coastal belt. Now there is need to educate and persuade farmers to opt for such practice which is less costly, eco friendly, easy to adopt with less labour requirements. The general attitude of the farmers now has been towards inorganic farming. To change their attitude towards sustainable agriculture, demonstration, training, group discussion and constant follow up programme would be much helpful.

Table 3.3: Nutrient content of green manure and green leaf manure (% on dry wt. basis)

Crop	N	P	K
Green Manure			
1. Sesbania aculeta	3.3	0.7	1.3
2. Crotalaria juncea	2.6	0.6	2.0
3. Tephrosia speciosa	2.7	0.5	2.2
4. Phaseolus trilobus	2.1	0.5	-
Green Leaf manure			
1. Pongamia glabra	3.2	0.3	1.3
2. Glyricida maculleata	2.9	0.5	2.8
3. Azadirachta indica	2.8	0.3	0.4
4. Calatropis gigantea	2.1	0.7	3.6

V. Ecological Pest Control

Ecological pest control is a practice of sustainable agriculture. Man-made chemicals used for pest and disease control are harmful as it (i) kills useful insects (ii) bad for human health (iii) may stay in body of animals and environment (iv) chemical becomes resistant over a period of time (v) more expensive and (vi) provides short period solution.

Non-Chemical methods of pest control are preferred in sustainable agriculture. These methods may be three types (i) Cultural (ii) Mechanical or Physical and (iii) Biological.

1. Cultural Method of Pest Control

(i) **Sanitation:** It includes removal of or destruction of breeding refuses and overwintering sites of pest. Removal of diseased plants or pruning infested parts of the plant is important disease management process. To prevent disease and pests, all means carrying the pathogens are to be destroyed. This suppresses pest population.

(ii) **Tillage**: Tillage is the most popular means of controlling weeds. The frequency, depth, alternate plough put down the pest and weeds. With soil tillage other pests are also killed. Deep ploughing in summer controls weeds and pests.

(iii) **Soil Amendments:** Application of soil amendment would change the rhizospher environment by affecting porosity, aerotion, temperature, water holding capacity. Healthy plants can better fight than under nourished plants. Decomposition of amendment produces direct effect which control many plant diseases.

(iv) **Habitat diversification:** Provision of irrigation, development of short duration crop, photo-insensitive, high yielding varieties of crops have resulted in growing more number of crops in a year. This is how pests are controlled to great extent.

(v) **Crop Rotation:** Good crop rotation with non-host plants results in reduction of pest population. Several soil born diseases of cereal crops are successfully controlled by a crop rotation period of 2-3 years. Crop rotation is one of the most important measures for controlling plant parasitic nematodes and is economic. Rotation is most effective against soil pests like: white grubs, wire worm, etc.

(vi) **Trap Cropping:** The practice attracts to small plantings of crops which are then destroyed or sprayed with toxicants. It is effective for nematodes, parasitic weeds and insect pests.

(vii) **Intercropping:** Inter -row space is a potential place for weeds to grow and these can be put to better use by incorporating them into soil. Lower incidences of pests are found with legumes intercropped with maize, clover, spinach, beans, tomato, etc.

(viii) **Strip Farming:** Intervening strip of non-suitable crop prevent movement of insect pests from one strip of crop to another. This method is also practiced where adjacent strips share unspecialized natural enemies.

(ix) **Time of Planting:** Pest outbreaks occur at particular soil and climate conditions and planting can be adjusted that such outbreaks do not coincide with the susceptible stage of crop. Late sown crop in winter hasfewer problems of nematodes.

(ix) **Water management:** It is more effective in potato to control scab. Disease which are favored by furrow, high moisture at soil surface like damping off, collar rot can be minimized by planting crops in ridges or raised beds. Irrigation avoids such diseases. Irrigation also suppress the weed population if flooded up to 20-30 cms.

(x) **Crop Competition:** Weed and crop seeds germinate at same time and start competing with each other for water, nutrients and space. In weed management vigorous growing crops, close row spacing, higher seed rate, proper time and method of fertilizer application are quite important.

Table 3.4: Crop competition and weed suppression

Crops Cereals	Suppression (%)	Crops Non-Cereals	Suppression (%)
1. Maize	92.00	1. Cow pea	88.00
2. Pearl millet	88.00	2. Ground nut	62.00
3. Sorghum	81.00	3. Castor	51.00
4. Set aria	73.00	4. Pigeon pea	54.00

(*Source:* Sharma, A.K., A Handbook of Organic Farming, Agrobios p.303)

Physical/Mechanical Control

1. **Manual Control**: Hand removal of weeds, pest attacked plants
2. **Burning:** Burning rice straw reduce incidence of stem rot of rice
3. **Solarization:** Covering soil surface with polythene sheet increase soil temperature is lethal to soil borne pathogens. Mulching period of 2-6 weeks with clear polythene sheet is effective to control many annual weeds.
4. **Flooding:** Flooding of field is a classical method to control banana wilt. Many plant parasitic nematodes are controlled by flooding methods.

Vermi-Compost: Earth worms are the best friends of soil and our farming system. Now emphasis is given for vermi-compost to improve the soil condition.

Vermi Composting is a method of making compost with the use of earth worms which generally live in soil, eat biomass and excrete it in digested form. About 1800 worms is ideal population for one square meter can feed 80 tones of humus per year. Earth Worms are one of the numerous ranges of burrowing organisms which improve soils. There are about 509 species of earth worms and that we need not depend on exotic species of earth worms. Earth worms are (i) indicators of fertility (ii) provide better plant growth (iii) help in effective utilization of soil ecosystem and (iv) minimize pollution hazards.

(i) Advantages of Vermi-Compost

1. Easy management
2. Commercialization
3. Recycle huge organic matter

(ii) Selection of Species: It depends on biological and ecological parameters like habitat, characteristics, distribution in soil, feed media, tropic function etc. The species like (i) Epiges (ii) Endoges and (iii) Aneciques can be used for vermi-compost.

Suitability of Species for vermi-compost

1. Tolerant to disease
2. Simple culturing practice
3. High consumption, digestion and assimilation rate
4. Wide adaptability
5. Feeding preference on wide range of organic matter
6. Produce huge cocoons with less hatching time
7. Fast maturing
8. Comparatively tolerance to other worms
9. Integrated Crop Management (ICM)

Integrated Crop Management is suggested by the scientists for making agriculture sustainable. ICM is adopted to fulfill the following objectives.

1. To provide adequate food and other agro products
2. To minimize consumption of non-renewable resources

3. To preserve quality of soil, water and air

4. To preserve bio-diversity

VI. Integrated Crop Management

It is a method of farming that balances the requirements of running a profitable business with responsibility and sensitivity to the environment. It is "Whole Farm Approach" and location specific that includes (i) the use of crop rotation (ii) appropriate cultivation techniques and (iii) careful choice of crop and varieties.

VII. Use of Indigenous Technical Knowledge (ITK)

Indigenous Technical Knowledge (ITK) or people wisdom or local or native knowledge has been the main practices in all walks of life of rural people and more so in case of human health, crop production, animal husbandry, fisheries or any component of livelihood system. The women are more inclined to ITK as they are practicing them from generation to generation. ITKs are preferred because (i) locally availability, (ii) eco friendly (iii) cheap (iv) easy to prepare (v) known to many and (vi) with no ill effect or residual effect. Women come forward first to use ITK than male members of the society as these remain within their reach. The cited ITKS are well known to women, majority of them use these and advice to others for which these have been mentioned in the chapter.

VIII. Climate Change and Sustainable Agriculture

Globally one of the most vital environmental challenges faced is the climate change. It is a growing crisis that has various implications right from economic to crop production. Climate change and agriculture are interrelated process so also women living on agriculture. Global warming is projected to have significant impacts on conditions regulating production system. It is mostly rainfall and temperature that create problem in

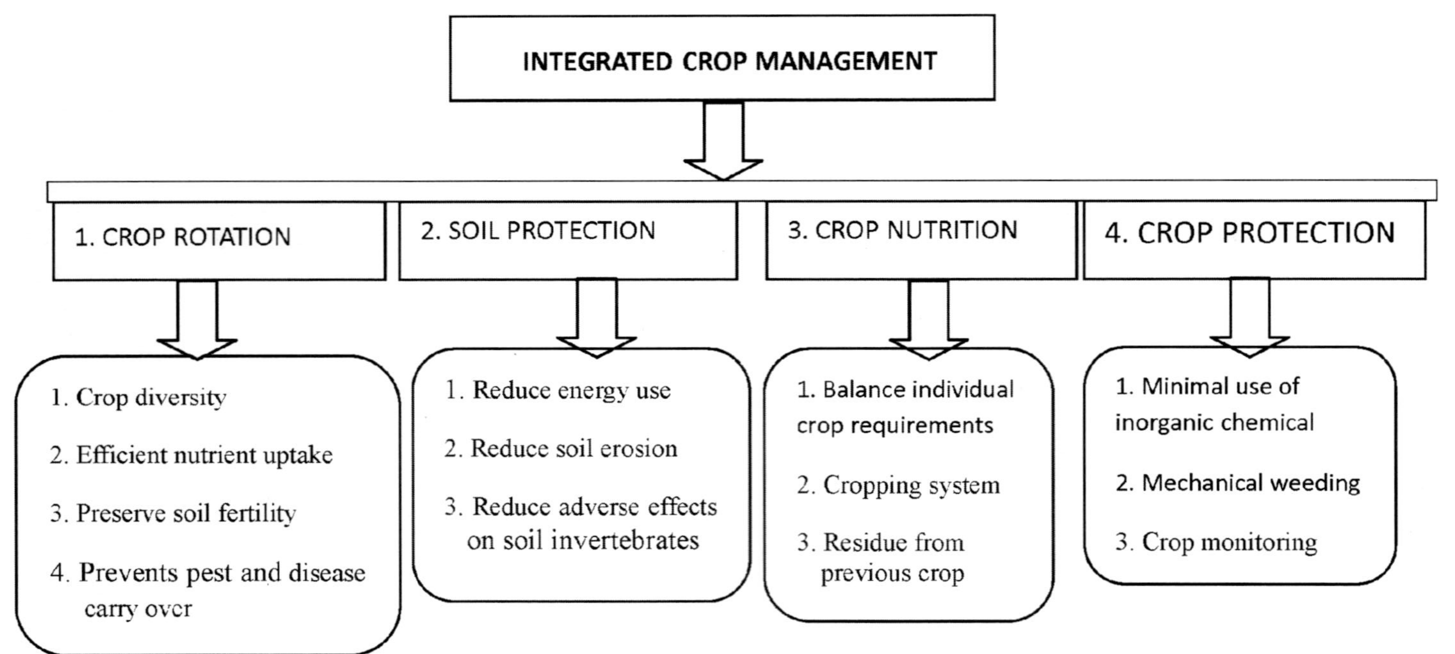

Fig. 3.1: Integrated Crop Management

Table 3.5: Women: The user of ITK

S.N.	ITK	Related to	Location
(A)	**AGRICULTURE**		
1.	Condensed clouds from North west direction ensure heavy rain.	Weather forecasting	Sambalpur
2.	Formation of misty circles around the moon at night ensures heavy rain shortly.	Weather forecasting	Khurda
3.	In the year if the leaf of Mahua Trees (flower of country liquor) falls uniformly around the trunk, it indicates normal rainfall for the year.	Weather forecasting	Koraput, Kalahandi tribal zones
4.	Bio-indicators of rainfall.	Rainfall indicators	Puri, Khurda
	(i) Sound of toad in continuation		
	(ii) Ants coming out of holes and move freely		
	(iii) Frequent deep of birds in sun		
	(iv) When cow coughs rain within 12 hours		
	(v) If parrot cuts mango tree in roundish way normal rain and in one side erratic heavy rainfall		
	(vi) When crab crawl to upland from river bank heavy rain leading to flood		
5.	The year in which sal plant *(shorea robusta)*Flowering is not uniform, i.e. flowering at basal part and no flowering at top drought immanent.	Drought indicator	Khurda, Daspalla, Nayagarh
6.	If east wind is stopped and west wind get started, there is assured rainfall.	Rainfall indicators	Khurda, Koraput
7.	Presence of ant hills in field indicates presence of underground water in less depth	Soil condition and underground water	Khurda, Ganjam
8.	Presence of ant hills in field indicates presence of underground water in less depth.	Presence of underg-round water	Koraput
9.	Plant features such as more number of branch, leave, long internodes indicate presence of water at low depth.	Depth of water in soil	Ganjam

Contd.

S.N.	ITK	Related to	Location
10.	Placement of bamboo stick in paddy field placed vertically and so far it does not fall, it indicate stress condition requiring irrigation in priority.	Need of irrigation	Ganjam, Sambalpur
11.	Control of soil erosion through composting land before rainy season, cultivation of deep rooted crops and plugging of gullies by stone.	Checking of soil erosion	Koraput
12.	Moisture and weed management covering rice straw to check evaporation loss.	Conservation of soil moisture	Koraput
13.	Sowing of small sized seeds (green vegetable, millets) mixed with ash, broken rice of soil particles.	Seed rate and sowing	Sambalpur
14.	Treating seeds with cow dung, honey and butter (2 part cow dung, 1 part money and 0.5 part butter.	Seed treatment	Khurda, Koraput
15.	Spray of urea solution to control weeds in sugarcane crop.	Weed control in sugar cane	Ganjam
16.	Control of termites in sugar cane field by applying lime and salt.	Termite control in sugar cane	Ganjam
17.	Control of stored grains pest in pulses by using dry chilli.	Control of stored grain pest	Ganjam
18.	Mustard oil and sesamum oil used to control stored grains of pulses.	Control of stored grain pest	Cuttack, Puri
19.	Application of cloves (one or two) controls ants in sugar and sugar packets.	Ant control	Ganjam
20.	Banana as indicator plant to irrigate sugarcane crop(when banana plant shows symptom of wilting).	Irrigation in sugar cane	Ganjam
21.	Management of Gundhi (harmful green algae) in paddy field by use of Karada leaves (*Cleistanthus collinus*).	Pest control in paddy	Nayagarh
22.	Control of Gandhi bug by spray of garlic and tobacco leave extract with washing powder.	Pest control in paddy	Puri

Contd.

S.N.	ITK	Related to	Location
23.	Control of case worm (*Nymphula depunctalus*)by leaves of sal.	Pest control in paddy	Bolangir
24.	Planting of wild saccharum spontaneum in paddy field for control of leaf folders.	Pest control in paddy	Ganjam
25.	Control of gull fly in rice by use of parso / prsu leaves.	Pest control in paddy	'Sambalpur
26.	Management of yellow stem borer in paddy by use of parasi (*Cleistanthus collinus*).	Pest control in paddy	Mayurbhanja
27.	Control of khaira disease in paddy by application of lime and cow dung.	Pest control in paddy	Cuttack
28.	Vegetable insects pests are controlled by application of animal urine and dusting cow dung ash.	Pest control in vegetables	Puri
29.	Control of insect pests of cucurbit crops by spray of cow urine in tobacco soaked water.	Vegetable pest control	Dhenkanal
30.	Control of shoot and fruit borer in brinjal by compost made of kochila and cow dung.	Vegetable pest control	Ganjam
31.	Control of shoot and fruit borer in vegetables by soaked water with soap.	Vegetable pest control	Bolangir
32.	Sprouting of yam is achieved by application of cow dung slurry.	Germination of vegetables	Ganjam
33.	Germination of vegetable seed is done by application of butter and milk. vegetables.	Germination of	Puri, Cuttack
34.	Enhancement of onion productivity by applying 75 kg fresh cassia leaves mixed with 125 kg cow dung when applied to onion crop yield increase by 25-30%.	Increase of productivity	Tamil Nadu
35.	Pest management of tomato by applying extract of leaves of Cynodon dactylon (duba ghas).	Pest control	Tamil Nadu
36.	Dark brown colored soil are fertile, light yellow as medium fertility and red as soil infertile.	Soil fertility	Khurda

Contd.

S.N.	ITK	Related to	Location
37.	Ant hills are always found in infertile land where as earth worm presence indicates good fertility status.	Soil fertility	Sambal pur
38.	Ash when mixed with phosphate fertilizer and applied in chilly crop the efficiency of fertilizer increases.	Soil fertility	Ganjam
39.	Sesamum oil cake when placed in vegetable field, increases female flower leading to enhancement of production.	Increase of female flowers	Ganjam, Koraput
40.	Application of karla and kochilla twigs in paddy field attract spiders and increases its population which control caseworm.	Pest control	Sambalpur, Ganjam
41.	Placement of Kochilla twigs in bundles consisting of 10-12 in number, control crop from stem borer and leaf folder attack.	Pest control	Ganjam, Koraput
42.	Spraying of fresh cow dung in paddy controls BLB.	Pest control	Ganjam, Sambalpur
43.	Kerosene mixed with sand (1lit kerosene with 10 kg sand) control case worm in paddy.	Pest control	Sambalpur
44.	About 250-300 gms of young fresh bamboo sprouts are soaked overnight and next day sprouts are removed and water is sprayed in paddy field which act as repellant for many insects.	Pest control	Ganjam
45.	Rice husk mixed with horse gram (200gms) + (10gm) kept in air tight container save stored grains.	Control of stored grain pests	Sambalpur
46.	Application of neem leaf: Rice grain spread over a piece of cloth under hot sun, 100 gms of neem leaf and paddy grain up to 50 kg are allowed to sundry will control rice weevil.	Control of stored grain pest	Sambalpur
47.	Calf urine with turmeric powder (I lit urine + 50 gm turmeric power) can control burning effect of young paddy seedlings.	Saving of paddy seedlings	Khurda

Contd.

S.N.	ITK	Related to	Location
(B)	**LIVESTOCK MANAGEMENT**		
48.	Application of pest made from crushed tamarind leaves, turmeric and onion juice on the injuries of cattle.	Injury treatment of cattle	Khurda
49.	(i) Using Karia twigs with mud and the affected animals are made to walk on the mud. (ii) Application of coconut outer cover oil and bamboo node oil on affected parts. (iii) Making the affected animals walk on sand bed and then applying of neem oil and kerosene on affected parts. (iv) Making the affected animals walk on sand bed and then applying of salt and mustard oil on wounds.	Control of FMD in cattle	Sambalpur
50.	Use of harida (*Terminalia chebula*) and bahada *Terminalia ballarica*) in cattle.	Treatment of affected parts of hooves and mouth	Khurda, Ganjam
51.	Swelling and hardening of udder, oozing out of white mucus from vagina and humping on other animals.	Symptoms of animal heat period	Khurda, Ganjam, Sambalpur
52.	Feeding the animals with plant parts of jute or bamboo / feeding a mixture of wheat, mahua and common salt / feeding fenugreek powder with jaggery.	Animal come to heat	Khurda, Sambalpur
53.	Feeding the animals with satavari, garlic and mustard.	Easy birth	Khurda, Sambalpur
54.	Feeding animals with mahua. turmeric, grains and edible oils.	Proper growth of calf and cattle	Khurda, Ganjam
55.	Powder of satavari plant roots mixed with jaggery fed to milch cattle.	More milk production	Sambalpur, Ganjam

Contd.

S.N.	ITK	Related to	Location
56.	Application of linseed and lime water on burnt parts of cattle.	Treatment of fire burnt	Khurda
57.	Banding of red hot iron rods on the skin surrounding the swelling portions in animals.	Treatment of black quarter	Khurda
58.	Application of paste made from crushed leaves of wild onion (*Eurginea indica*) on udder of animals.	Treatment of mastitis	Khurda
59.	Feeding of a mixture of mustard oil and dry ginger and crushed ajawain seeds to suffering animals.	Treatment of gastritis	Khurda
60.	Feeding a mixture made from crushed powder of turmeric, hemp and ajawain seeds with curd.	Indigestion treatment of cattle	Khurda
61.	Use of ajawain, jaggery and mustard oil with cattle.	Overcome constipation	Khurda
62.	Applying a paste made from crushed mustard seed powder and water on mouth of cattle.	Treatment against cold and cough	Ganjam
63.	Feeding cattle 'uturuni' (local plant) root with jaggery.	Treatment of cold	Ganjam
64.	Apply of grounded and boiled green neem leaves on wounds of animals.	Cure of wounds	Khurda, Ganjam
65.	Feeding the cattle with roots of 'koilikhia' plant with jaggery.	Cure of digestive disorders	Ganjam
66.	Apply of grounded wood coal on wounds of animals.	Cure of wounds	Ganjam, Khurda
67.	Feeding crushed roots of 'Bichhuati' plant (local weed) with jaggery to cattle.	Curing skin infection	Khurda
68.	Feeding the animals with crushed leaves of 'Dhanatwari' (local medicinal plant) with jaggery.	Cure of diarrohea	Ganjam
69.	Pressing 'Sunari' pods over swollen throat of cattle.	Control of throat swellings	Sambalpur
70.	Mixing rice grains from un-boiled paddy and tied over the broken horns of cattle.	Treatment of broken horns	Sambalpur

Contd.

S.N.	ITK	Related to	Location
71.	Feeding the cattle with boiled neem bark water.	Control of worms	Khurda, Sambalpur
72.	Use of paste made from stem of 'Hada Bhanga' and tied at the place of fracture of bones.	Setting of bones of animals	Khurda, Sambalpur
73.	Massaging affected part of cattle with mustard oil.	Control of neck inflammation	Sambalpur
74.	Feeding cattle the mixture composed of 'Begunia' leaves with spider thread and salt.	Control of loose motion	Sambalpur
75.	Feeding lemon grass to cattle.	(i) Control of loose motion (ii) Enhances milk production	Khurda, Sambalpur
76.	Feeding cattle the mixture composed of 'Ajawain', rock salt, dried ginger and jaggery.	Care of fever and indigestion	Sambalpur
77.	(i) Feeding neem leaves paste(ii) Feeding 'Baidanka' fruit, curd or tamarind.	De-worming of calf	Koraput
78.	Use of roots of 'Vaira' plant on animal body.	Eradication of lice	Koraput
79.	Touching the cut portion of 'Bhalia' seed above the eyelid of animals.	Curing wound in eye	Koraput
80.	Application of juice or milk of 'Arakha' (calotropis) on the wound to act as anti-septic.	Wound treatment of cattle	Koraput
81.	Exposing cattle with 'Jhuna' smoke in cowshed.	Treatment of neck swelling	Khurda, Koraput
82.	Rubbing the swollen portion of tongue in animals with help of stone followed by rubbing with old tamarind fruit and dry chilli powder.	Treatment of tongue swelling	Koraput
83.	Feeding a well mixture made from ragi gruel + tamarind fruit.	Control of diarrhea in animals	Ganjam, Koraput

Contd.

S.N.	ITK	Related to	Location
84.	(i) Feeding the mixture made from fresh 'Rukuni' bark, horse grain seeds and husk.(ii) Feeding grounded 'Rajdumuri' fruits.	Increase of milk yield of milch cows	Koraput
85.	Feeding onion juice / feeding country liquor and opium to the affected poultry birds.	Control of Ranikhet disease	Koraput
86.	Exposing poultry birds with smoke of burnt 'Begunia' leaves.	Eradication of ticks/ lice	Khurda, Koraput
87.	Use of termite ant hill soil in goats and sheep.	Treatment of wounds	Koraput
(C)	**FISHERIES**		
88.	Sprinkling fresh lime solution in pond.	Fish free from diseases	Ganjam
89.	Allowing buffaloes to swim in the fish pond.	Supply of oxygen to fish in fresh	Ganjam
90.	Putting banana leaves and twigs of caster plant in fish pond as feed.	Quick growth and maturation of fish	Ganjam
91.	Putting fresh cow dung in fish pond to help in growth of plants.	Plants as feed for fish	Ganjam
92.	Sun drying of small fishes over the sand bed of sea coast.	Safe storage of dry fish for 8-10 days	Ganjam
93.	Sun drying of fish after salt solution.	Preserves dry fish for 2-3 months	Ganjam
94.	Use of neem cake / mahua cake / karanja cake in pond for creating de-oxidation.	Easy catching of fish	Sambalpur
(D)	**POST HARVEST TECHNOLOGY**		
95.	Use of 'Mushala' for de-husking of rice, ragi, niger and 'karanja' seeds.	Post-harvest processing	Koraput
96.	Mixing of ash with pulse grains.	Protection from insects and pests	Khurda, Sambalpur

Contd.

S.N.	ITK	Related to	Location
97.	Use of 'Moringa' leaves and crushed camphor with pulse grains.	Safe storage of pulses	Khurda
98.	Mixing cow dung with vegetable seeds before storage.	Effective storage of veg. seeds	Khurda, Sambalpur
99.	Use of neem / 'begunia' leaves with grams.	Repellent effect helps in storing	Khurda, Sambalpur
100.	Adding fried rice grain with mung bean.	Protection from insects and pests	Sambalpur
101.	Putting laterite stone inside paddy bag.	Control of stored grain pests and insects	Sambalpur
102.	Putting sand and ginger in layers.	Safe storage of ginger	Khurda, Ganjam, Sambalpur
103.	Hanging seeds of gourds over smoke chulla.	Safe storage of seeds	Sambalpur
104.	Sun drying of small pieces of various vegetables (brinjal / cabbage / tomato).	Preserving vegetables	Khurda, Sambalpur
105.	Putting neem and 'karanja' leaves in rice bag.	Storage of rice	Khurda, Ganjam, Sambalpur
106.	Mixing crushed lime powder with rice in bag.	Storage of rice	Ganjam
107.	Turmeric powder and red chilli are tied and used for rice grains.	Protects from weevils	Khurda, Ganjam
108.	Mixing common salt with rice grain.	Protects from weevils	Ganjam
109.	Mixing red soil and ash with arhar seeds before storage.	Keeps free from insect pests	Khurda, Ganjam
110.	Green banana inside sand layer.	Safe storage without damage for one week	Ganjam
111.	Keeping pulse grains inside mud and sand layer of an earthen pot.	Safe storage up to 6-8 months	Ganjam

Contd.

S.N.	ITK	Related to	Location
112.	Keeping colocassia plants as such in the field for seed purpose.	Stored safely by natural condition	Ganjam
113.	Mixing arhar with ash and keeping in earthen pot.	Storage purpose	Koraput
114.	Use of structure made up from 'Siali' leaves for storage of salt.	Preservation of salt	Koraput
115.	Mixing rice husk with horsegram.	Avoids storage pests	Sambalpur
(E)	**PLANT PROTECTION**		
116.	Application of 'Kochila' plants and twigs of 'Karia' in rice field.	Attracts spiders which control case worm	Ganjam, Sambalpur
117.	Burning cycle tyre in paddy fields.	Control of Gundhi bugs	Sambalpur
118.	Burning stubbles in the bonds of paddy fields.	Improves soil fertility	Khurda, Sambalpur
119.	Sprinkling leaf extracts of lemon grass and *Cassia occidentalis*	Controlling cater pillars	Ganjam
120.	Plantation of garlic as a border row in chilli field	Acts as repellent	Ganjam
121.	Planting 'Tulusi' (*Ocimum sanctum*) in brinjal fields.	Protecting from insect pests and diseases	Ganjam
122.	Lady's finger planted in brinjal and tomato field.	Protects nematodes	Ganjam
123.	Banana planted in sugarcane fields.	Protects from red rot diseases	Ganjam
124.	Putting 'Kolathia' plants in paddy fields.	Controlling stem borer and leaf folder attack	Ganjam, Koraput
125.	Sprinkling wood ash over country bean plants.	Protects from aphids	Ganjam
126.	Sprinkling the solution made from crushed seeds of custard apple (*Annonas squamosa*).	Controlling diamond backmoth	Ganjam

Contd.

S.N.	ITK	Related to	Location
127.	Application of 'karanja' and neem cakes in brinjal plants.	Controlling fruit and shoot borer	Sambalpur
128.	Application of 'Mahua' oil cake in paddy fields.	Protects stem borer	Sambalpur
129.	Spraying of cow dung slurry over the crops.	Controls B. L. B.	Ganjam, Sambalpur
130.	Application of neem cake in paddy nursery bed.	Acts against BPH	Ganjam, Sambalpur
131.	Dusting ash over the stems and leaves of gourds.	Helps in aphis control	Sambalpur, Ganjam, Koraput
132.	Use of tin containers in sun flower and maize fields.	Bird searing	Khurda, Sambalpur
133.	Application of maned cake (made from rice grains) in paddy fields.	Controls all insect pests	Sambalpur, Ganjam
134.	Putting one stomp by the side of the rat hole.	Rat killing	Sambalpur, Ganjam
135.	Application of neem cake in lemon and coconut.	Controlling white ant	Khurda, Sambalpur
136.	Planting marigold plants as border row crop.	Helps in nematode control	Ganjam, Sambalpur
137.	Spraying of cow dung slurry in ginger.	Case worm control	Sambalpur
138.	Application of churn curd in lemon.	Controls fruit drop	Sambalpur
139.	Spraying calf dung slurry in rice seedlings.	Reduction of burning effect in seedling	Khurda, Sambalpur, Ganjam

Contd.

S.N.	ITK	Related to	Location
140.	Application of calf urine and turmeric powder in rie seedlings.	Minimizes burning symptoms of seedlings	Khurda, Sambalpur
141.	Use of young bamboo sprouts in paddy field.	Acts as repellent for insect pests	Koraput
142.	Use of 'Salap' tree (a palm) in paddy crop.	Controls leaf folder and case worm	Koraput
143.	Use of 'Suneli' (local tree) branch in paddy.	Rats control	Koraput
144.	Dripping cow urine or goat urine into 'Salap' tree.	Kills the bettle	Koraput
145.	Application of soap and kerosene in vegetable.	Kills micro organisms	Khurda, Koraput
146.	Use of 'Vaira' roots in vegetable crops.	Effective control of insect pests	Koraput

(*Sources:* Collected from literatures, local people and key communicators of study areas)

existing farming system. Temperature, carbon dioxide, precipitation and interaction of these factors decide fate of production of agriculture. The overall impact of crop production depends balancing of these factors. Climate change has visible impacts not only on the crop production but also on human health, water supply and energy. Destabilization in rainfall and temperature shows impact on occurrence of drought, flood and cyclone.

(i) Causes of climate change: The major causes of climate changes are release of carbon dioxide and green house gases. There are several factors which are directly related to climate change.

1. Increase in average mean temperature
2. Change in rainfall pattern and amount
3. Rising of atmospheric carbon dioxide
4. Pollution level
5. Variability extent

Table 3.6: Sources of green house gas emission

Sources	Percentage
1. Industrial processes	16.8
2. Power station	21.3
3. Water disposal and treatment	3.4
4. Land use and biomass burning	10.0
5. Residential, commercial and other sources	10.3
6. Fossil, fuel, retrieval processing and distribution	11.3
7. Agricultural and by product	12.5
8. Transport fuel	14.0

Table 3.7: World's top emitters (2005)

	Countries'	Carbon Dioxide in billion tones(GI)
1.	USA	5.96
2.	China	5.32
3.	Russia	1.70
4.	Japan	1.23
5.	India	1.17
6.	Germany	0.84
7.	Canada	0.63
8.	Britain	0.58
9.	South Korea	0.50
10.	Italy	0.47

Source: Netherland's Environment Assessment Agency, 2005

Unprecedented climate change on a global scale during past decade resulting from green house effect out of anthropogenic reasons has serious implications on the natural ecosystem and socio-economic conditions. A rise in Earth's means temperature could significantly modify the quantity and pattern of rainfall which in turn may boost occurrence and concentration of severe climatic events such as flood, famines, heat waves, cyclones and tornadoes.

Box No. 3.3 Impact of climate on agriculture

1. Increase in rate of migration
2. Change in occupational structure
3. Shortage of labour
4. Increase in requirements of labour per unit area of operation
5. Increase in social mobility
6. Increase contact with market
7. Change in customer preference for agricultural produce
8. Down rating farming as remunerative profession
9. Increase in group pressure on policy makers for subsidy
10. Social disturbances

Future green house emission depends on the development of pathways driven by economic, demographic, land use, agricultural and energy drivers. The interaction among these key driving force is a very complex and has regional specificity. India with 2.5% of the total global land area supports population of about 11 billion. Increased population pressure and limited resources may affect agriculture, water resources or energy consumption. The need for greater promotion of use of renewable sources of energy such as wind, water and solar energy need no emphasis. Agriculture in India contributes about 25% of the domestic products. Climate change makes agriculture vulnerable because it depends heavily on weather.

Climate change and its impact on agriculture: Different studies indicate that there will be substantial reduction in water availability in large part of Northern India. According to the findings of IPCC (Inter Governmental Panel on Climate Change) the consequence of environmental changes in South-East include an increased risk of floods and droughts in many regions, decreased agricultural productivity, adverse impact on fisheries and adverse impact on many ecological systems. Wheat yield would be reduced by 5-10% with every increase of one degree centigrade and overall yields could be decreased up to 30% in South-Asia by mid 21st Century. India could experience 40% decline in agricultural productivity by 2080. The way the climate change is taking place; there will be much more if we do not take care from the beginning. The amount of climate change will depend on:

(i) Increase or decrease or same in gashouse emission and particularly aerosol

(ii) Response of temperature, precipitation and sea level to changes in gashouse emission

(iii) How much climate varies and its internal variability

In the long run, the climate change will affect our agriculture like (i) productivity in terms of quantity and quality (ii) use of inputs such as irrigation, herbicides, insecticides and fertilizer (iii) frequency and intensity of soil drainage (nitrogen leaching, soil erosion etc. (iv) rural space like loss and gain in cultivated land, land distribution and water availability in different soils and (v) adaptation of crop and varieties.

Strategy to mitigate climate change in agriculture includes various parameters. These are:

1. Use of fuel –efficient pump sets
2. Rationalization of power tariffs
3. Better cultivar practices which will help reducing N_2O emission
4. Management of rice production through judicious use of organic manure, fertilizers, irrigation water, nutrification, fertilizer placement and their scheduling
5. Improved management of livestock population specially ruminants and its diets
6. Increase soil organic carbon through minimal tillage and residue management
7. Improve energy use efficiency in agriculture through better designs of machinery and by resource conservation practices
8. Change land use pattern by increasing area under bio-fuel, agro forestry but not at the cost of food production
9. Establishment of climate information management centers
10. Build climate risk assessment for long lived infrastructure projects
11. Emphasis on research and extension support to provide sustainable modes of dry land farming

1. Direct effect on Physiology and Morphology of crops: Women sample (N=240) have expressed their realization about climate change in their farming activities. The realization is expressed in terms of deviation in sowing time of crops, transplanting, resistance to pest and diseases, harvesting time and different cultural practices.

Table 3.8: Opinion of farm women about climate change N=240

Observation	Frequency	Percentage
1. Late sowing	180	75.00
2. Late transplanting	165	68.75
3. Growth of weeds	98	40.83
4. Late harvesting	176	73.33
5. Unable to take second crop	82	34.16
6. Problems in drying produce	120	50.00
7. Small sized plants, leaves and grains	81	33.75
8. More disease and pest	109	45.41
9. Wilting of seedlings	121	50.41
10. Shrinkage area under pulse and oil seeds	185	77.08

2. Indirect Effect: Climate and rural living are interrelated and interwoven. All the occupational activities in the village depend on rainfall and temperature. There are certain areas where effect of climate change directly realized and in some cases it is indirect. The opinion of sample women about the indirect impact of climate change was ascertained.

Table 3.9: Opinion of farm women about status and fertility of soil N=240

	Observation	Frequency	Percentage
1.	Low soil fertility	76	31.67
2.	Cracking of soil	82	34.16
3.	Moisture deficiency	125	52.08
4.	Absence of residual moisture	139	57.91
5.	Flood	65	27.08
6.	Root rotting of cereals	87	36.25
7.	More of mites and white ants	90	37.50
8.	Seedling burning	72	30.00
9.	Killing of sprouted seeds	69	28.75
10.	Low yield	72	30.00

3. Effect of Climate change on socio-economic condition: In Odisha, the living style of the people is as per climatic condition. All the socio-cultural functions are observed depending on seasons like, summer, winter and rainy seasons. It has linking with growing of crops and their harvesting time. It is more so in case of tribes who follow season's climatic conditions. The changes in socio-economic condition due to change in climate are reported.

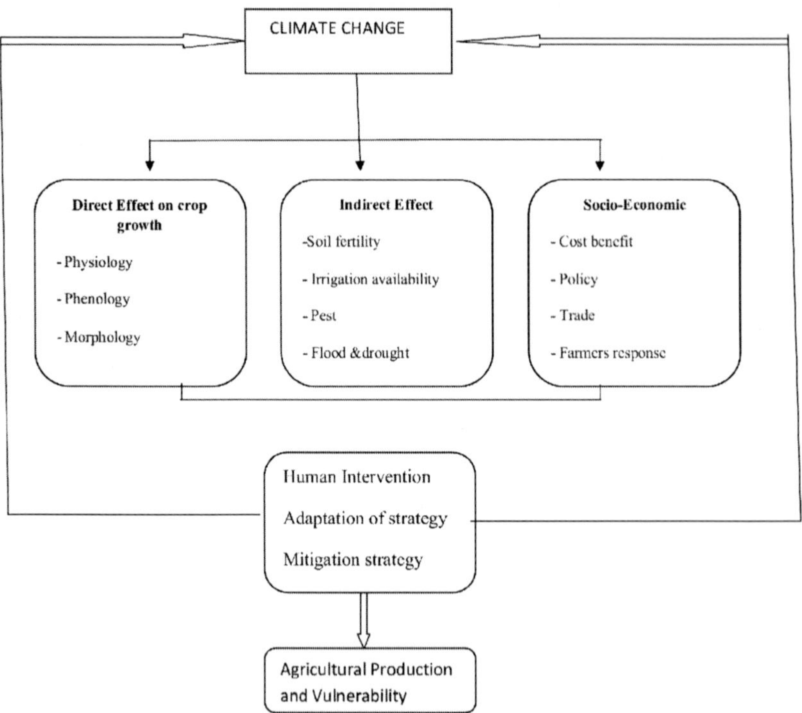

Ref: Recent Developments in Organic Farming, OUAT 2011

Fig. 3:2: Climate Change

Impact of Global Warming

1. The first and foremost impact of global warming is the rise of sea level. The mean sea level rise is due to two reasons. Firstly the atmosphere warms, the surface of the ocean warm as well expanding its volume then rising sea level. The second is polar ice glacier. Ice are melting due to increase in temperature. Worldwide sea level rose to 4 to 8 inches during the 20th century and it will rise to 37 inches in 21st century. According to NASA, the polar ie cap is now melting at the alarming rate of 9% per decade and Arctic ice thickness has decreased 40% since the 1960s. The changes in sea level will complicate life in many coastal regions. A 40 inches rise could submerge 6% of Netherland, 17.5% of Bangladesh. Densely populated coastal area of Florida, Louisiana would become uninhabitable. In India, Andaman Nicobar, Goa, Mumbai and some parts of West Bengal will be submerged under sea.

2. Climate Change is another major impact of Global Warming. It will bring about major changes in water distribution and have impact on water resources. Most glaciers will shrink and some small glaciers will disappear; water for irrigation will fall short due to high evaporation and due to high temperature. Similarly rain fall pattern will change. There will be more floods due to intense monsoon.

3. Global warming and environment pollution: Due to Global Warming pollution takes place in several ways. According to some estimates, the rice cultivation in the world is responsible for 20% methane being added to atmosphere and the coal mining accounts for 6% of methane. Similarly the deforestation is responsible for 20% of the carbon dioxide gas being added to the atmosphere and industrialization is adding 25% chlorofluorocarbon to the atmosphere.

Table 3.10: Percentage increase of concentration of green house gases and their warming capacity

	Name of green house gases	Average increase rate per year	Warming capacity	Contribution to global warming (%)
1.	Carbon dioxide	0.5	1	35
2.	Methane	1.0	36	20
3.	Nitrous oxide	0.3	140	5
4.	Chlorofluorocarbon	0.4	17000	12
5.	Ozone(O3)	0.5	430	2
6.	Hydroflorocarbon	0.4	14600	6

In addition, we can see the total carbon emission by major polluting countries.

Table 3.11: Total carbon dioxide emission by some countries from 1950-2002

	Countries	Emission (billion tons)
1.	USA	186.1
2.	EU	127.8
3.	Russia	68.4
4.	China	57.6
5.	Japan	31.2
6.	India	15.5
7.	Mexico	7.8
8.	Australia	7.6

CHAPTER 4

Adoption Behaviour of Women Farmers

In rural development, the sole aim is to transfer the technology for better living of people. The process by which the technology is transferred to users is called **Transfer of Technology.** Transfer of technology falls in the domain of adoption and diffusion. Therefore, it is necessary to understand adoption and its related issues. Adoption of sustainable agricultural practices is relatively difficult as people do not find newness in it. The transfer of technology is explained in terms of technology movement from place of generation to the hands of the users.

To provide an understanding of what is Transfer of Technology, we can examine the different definitions of it. In a simple meaning, the transfer of technology ends only after its use and satisfaction of the users.

I. Definitions of Transfer of Technology

1. Transfer of Technology is that part of extension strategy which covers communication, adoption, diffusion, advertisement and program decisions to make farming communities better than existing situation.

2. Transfer of technology is the process of bringing out tested and usable technologies from the researchers to the field of the users in agriculture and allied areas.
3. Transfer of technology not only indicates the transfer of technology to the users but also acceptance and use as per recommendation.
4. Transfer of technology denotes dissemination and translation of technology into action.
5. Transfer of technology is the process by which well tested and location specific technologies are transferred through appropriate channel of communication to the users and enter into farming system providing expected benefits to the users.
6. Transfer of Technology is also defined as dissemination of technical know-how to the users irrespective of sources for higher out-put.

Whatever the definitions of technology transfer may be, the theme is technology has to be moved from point of generation to the users. It involves a complex process where multidisciplinary approach is a must.

To illustrate the linkage between technology generation, technology transfer and technology used, the model is cited below.

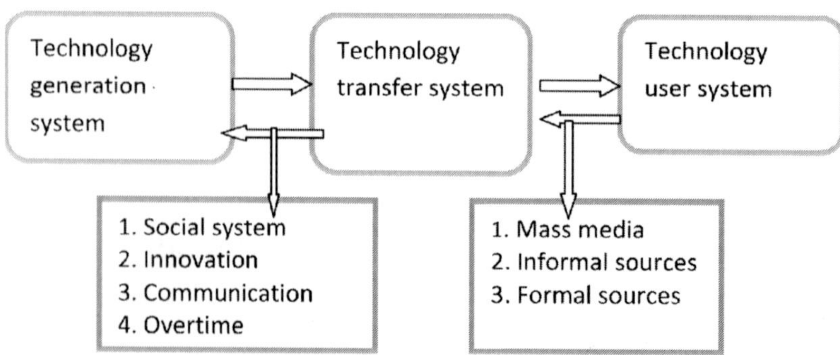

The process of transfer of technology brings close linkage between extension system, research system and users system. In case of transfer of sustainable practices in agriculture that too among women the same system operates with little deviation.

i) **Technology generation system:** An effective technology generation system should be based and designed as per field problems or needs of farmers along with constraints in field operation. It is the realization by field functionaries that all technologies are not adopted by the farmers for the simple reason of non-requirement and non-applicability. The desirable traits of technology decide the acceptance and adoption of technology. Research has enlisted 12 important desirable traits of technologies that accelerate the rate of adoption. In the context of sustainable agricultural practices these may be examined.

ii) **Technology Transfer System:** Technology transfer system is comparatively difficult as it deals with human beings who are never under influence of any body. The system

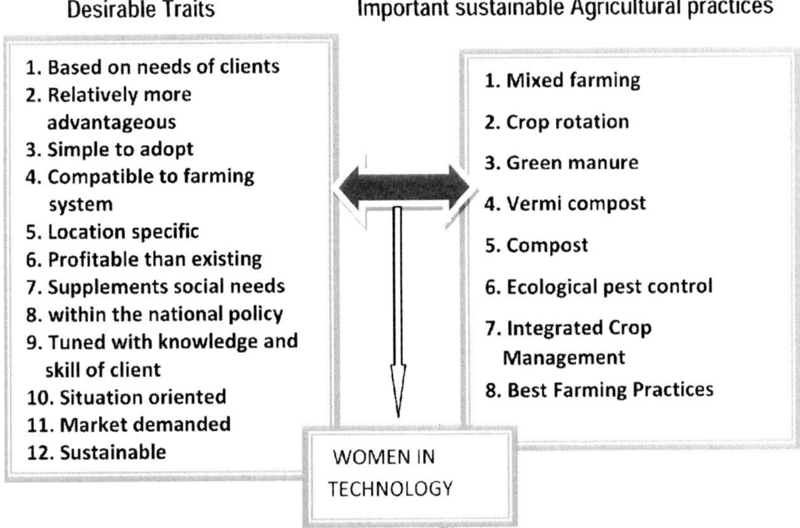

Fig. 4.1: Desirable Traits and Important Sustainable Agriculture Practices

involves motivation, persuasion which is dependent on decision making process. It has two major components, (i) Knowledge Transfer and (ii) Input transfer. Knowledge transfer is relatively difficult and time taking as it requires motivational force in the line of learning behavior. The input transfer requires financial involvement. But to put input into action is the domain of knowledge for which it always occupies top priority. The interrelationship of various factors associated with Technology Transfer System is illustrated below.

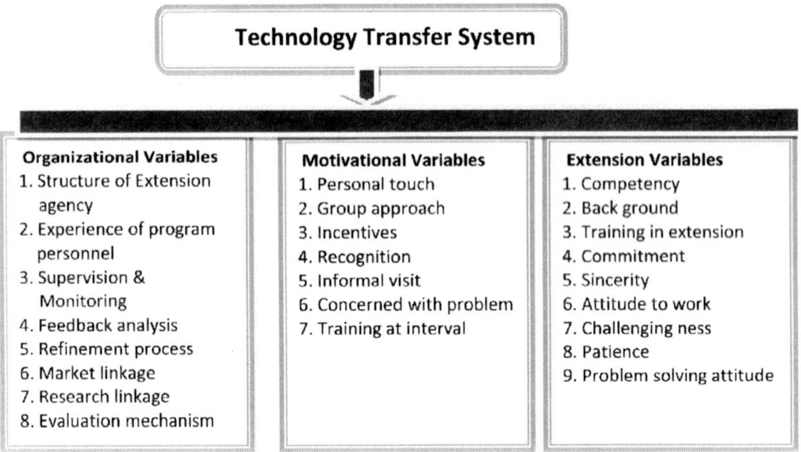

Fig. 4.2: Technology Transfer System

iii) **Technology User System**: Users are the most important factor in transfer of technology. In this case, women farmer need more care and approach to adopt sustainable agricultural practices. Whatever efforts organization or extension personnel take, the final decision maker is the consumer of technology. The farmers and their families are conditioned by certain factors which exert influence on rate of transfer of technology. Technology user system is regulated by personal, social, economic and situational factors.

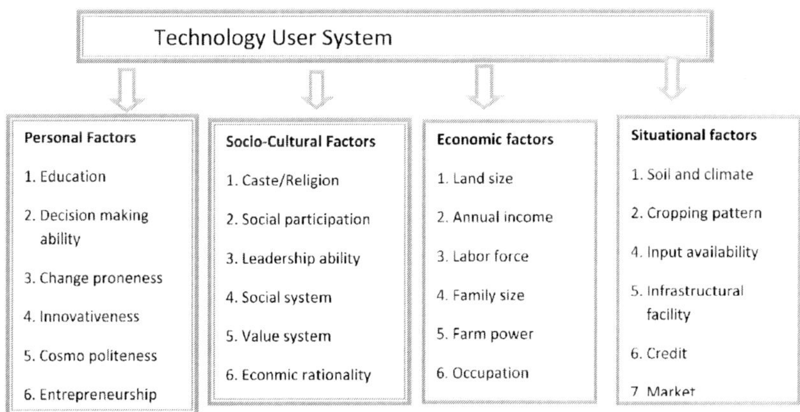

Fig. 4.3: Technology User System

Transfer of technology is a complex process and involves multiple factors. Transfer of technology involves technology generation, technology transfer and technology use with equal emphasis to all these elements.

In most of the developing countries, the subtropical and tropical climate the maintaining of organic matter becomes difficult under the traditional farming system which requires special efforts for sustainable agricultural practices. The reasons are (i) most of the arable soils contain organic carbon below threshold level and (ii) farming communities are resource poor to afford for costly inorganic fertilizers.

Lack of location specific technology to recycle organic wastes and lack of awareness to recycle organic wastes in agriculture are the main reasons for slow adoption even though it is the native technique for farmers which got lost during green revolution period. The harmful practices in the modern farming are:

1. Excess tilling
2. Excessive use of water
3. Use of inorganic fertilizers

4. Use of chemical pesticides
5. Absence or reduced natural allied farming
6. Use of varieties inconsistent with local factors

Organic practices are more environmentally sustainable than most conventional practices. The farmers are to be educated about the sustainable farming system, about the high potential areas. Marginal lands, organic agriculture often agronomically feasible includes positive microeconomic effects for most of the producers involved. Farmer knowledge, best solution for farm management, problem arises in field and potential solution is to come from the users of organic technologies. Farmer-back-to farmer model may be examined in case of adoption of sustainable practices in agriculture.

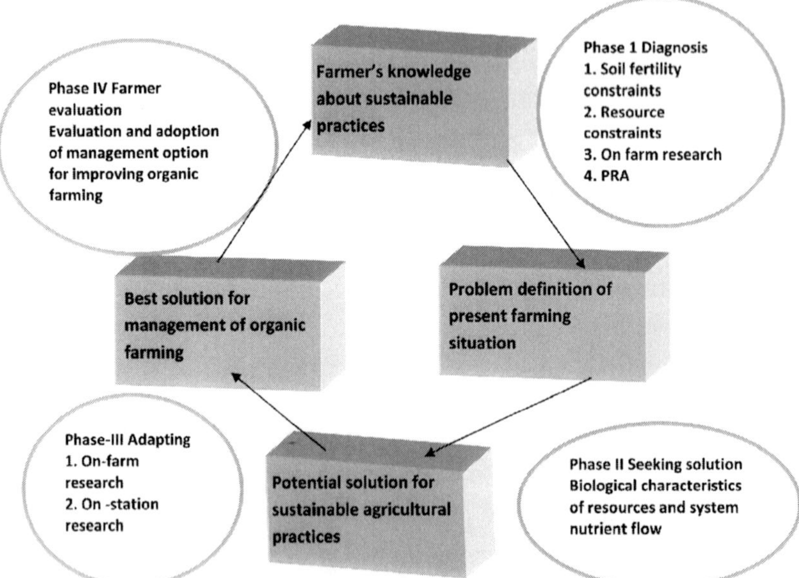

Fig. 4.4: Farmer's knowledge about sustainable practices
Source: Former-back to farmer approach to agriculture problem identification and solution, Rhodes and Booth, 1982.

II. Adoption Behaviour of Women Farmers

In rural development everybody is interested to see that transferred technology is put to action. When farmer or farm woman is faced with a technology they consider from different angles to take decision as whether to adopt it or not. Thus decision making process is very important in adoption of technology.

In case of sustainable agricultural practices our aim is that women are to to be educated to know, realize and adopt the following practices.

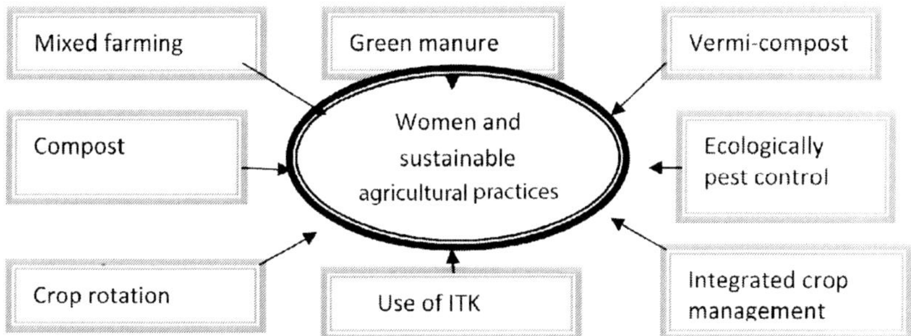

Fig. 4.5: Women and Sustainable Agricultural Practices

In the context of adoption, it is necessary to understand what is adoption and its dimensions. Adoption is defined as a decision to make full use of innovation whereas diffusion is the process by which the innovation is communicated through different channels over time among the members of a social system. Both the processes, i.e. adoption and diffusion are interrelated.

Issues in Adoption of Technology: There are certain established facts that the extension workers have to know while dealing with women for adoption of sustainable agricultural practices.

1. Decision to adopt usually takes time. It ranges from days to years as the decision depends on a variety of factors.
2. People pass through a series of stages before final adoption of the practices like awareness, interest, trial, evaluation and adoption. In case of women the movement from awareness to final decision takes longer time.

3. All people do not adopt at the same time with same intensity. Adoption depends on variety of factors like immediacy of needs and sources of information. Some sources are only useful at awareness stage while others in other stages. Therefore adoption of all farmers at a time never happens.

4. Final adoption is not permanent in nature. Adoption behavior of people changes with time and new innovations giving more satisfaction. Some people discontinue after sometimes while those who refuse at the beginning again adopt after sometimes.

5. People follow existing decision making process in case of adoption. The farmers take time to decide as adoption involves money, time and risk.

6. Individuals are important sources of information to influence others in adoption of technology.

7. Generally more experienced and knowledgeable persons act as advisers to others in a social system. In adoption of technology this type of information appears to be more effective than non-personal or impersonal sources to provide information to others.

8. Influential and innovator may not be same persons. Adopter adopt the practice after knowing its merit but innovators go on adding to the technology because of their creativeness.

9. Issues in adoption of technology are multidimensional. It is also dynamic in nature as it relates to market demand, choice of adopters and his family requirements.

10. Decision making process in family exerts appreciable pressure on adoption behavior of the family heads. With apprehension of profit, risk involvement and investment,

the decision makers face cross road problems. Human psychology is also to be considered in case of adoption issues.

III. Promoting Factors of Adoption of Technology

There are certain factors which accelerates rate of adoption among women of rural areas.

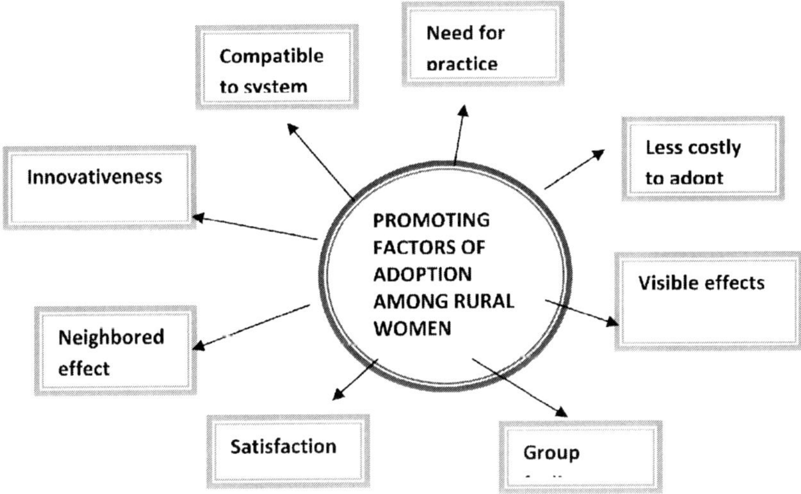

Fig. 4.6: Promoting Factors of Adoption Among Rural Women

Adoption Stages and Sources of Information: Farmers receive information from multiple sources. These sources are of different types in structure and function. When farmers move from one stage of adoption to another, these sources stimulate and encourage them to move. Again the farmers generally move from one stage to another stage. The stages are awareness, interest, evaluation, trial and adoption. At each stage the information requirement is different. The movement of farmers depends on capability of sources to stimulate them and win their psychology.

Man lives in society. Every moment he needs information to communicate. The process of communication has relevance with adoption of technology. The effectiveness of sources influencing the receivers of message is of greater importance in all steps of adoption process. The sources of information are classified according to their structure and function and these are of three types as mentioned below.

Table 4.1: Sources of information and stages of adoption process

Type of sources	Channels/sources
1. Mass media sources	TV, radio, film, newspaper, farm magazines, leaflets, bulletins
2. Formal sources	VAW (male and female), scientists, salesmen, traders, company representatives, any one working in formal organization in extension line
3. Informal sources	Friends, relatives, neighbour, progressive farmer, leader, influential key communicators

Researchers in extension domain have identified multistage in adoption process. The most acceptable stages are awareness, interest, evaluation, trial and adoption. These are at par with stages of learning process. The stages of adoption process and corresponding effective sources of information are presented in table below.

Table 4.2: Stages of adoption process and effective sources of information

Si. No.	Stages of adoption process	Sources/channel of information
1.	Awareness	Mass media
2.	Interest	Mass media, Informal sources
3.	Evaluation	Formal, Informal sources in mix
4.	Trial	Formal, Informal sources
5.	Adoption	Formal, sources of information

IV. Adopters and Stages of Adoption Process

The interest of change agents and sources of information is to help clients to move from awareness to final stage of adoption process. Such movement is conditioned by the factors like (i) interest of the client (ii) effective sources of information (iii) affordability (iv) technology and (v) other environmental situations. In a study conducted in Odisha taking cereal, oilseed and vegetable crops the movement of farmers from awareness to adoption stage was found out as depicted in following table.

Table 4.3: Movement of farmers from awareness to adoption stage N=240

States of adoption process	Cereals (n=200)		Oilseeds (n=180)		Vegetables (220)	
	f	%	f	%	f	%
1. Awareness	200	100.00	180	100.00	220	100.00
2. Interest	121	60.50	130	72.22	175	79.54
3. Evaluation	48	24.00	52	28.88	61	27.72
4. Trial	15	7.50	27	15.00	38	17.27
5. Adoption	12	6.00	18	10.00	23	10.45

The estimation was made out of the total sample from the stage of awareness although total sample were 240. At adoption stage the adopters are found within 11% of the total farmers made aware of. This really happens because a number of factors interplay to bring farmers to the stage of adoption. The movement of farmers in the stages of adoption varies from technology to technology depending on needs of the clients.

Box No. 4.1 Individual's look
1. Examining applicability of technology
2. Seeks advices
3. Discuss about technology
4. Takes decision to adopt or reject

Need and Adoption Behavior of Women: Adoption of sustainable practices depends on the need of the recipients of technology. Whether there is need for the practice or not is an important issue for consideration. People become interested when there is pressing need for the technology. Before creating awareness it is required to know the needs of the women farmers. There are number of methods for identification of needs. But while interacting with client one can easily know by observing whether need is at fore front or not. The phenomena that speak of needs are:

1. General discontentment without known cause and awareness
2. Long perception about specific discontentment
3. When cause and level of discontentment combine together the individual is ready for adoption of technology
4. Sometimes discontentment becomes cause of displacement

Attributes of Technology: Innovation is an idea perceived as new by individuals in his social system. Attributes are the characteristics which affect the rate of adoption. The technology is subjected to acceptance or rejection. It depends on satisfying quality of technology. These characteristics are known as attributes of technology. Sustainable agricultural practices are also technologies but not new in farming environment. Rather these are forgotten by the farmers. Attempt to bring them back to system may be considered on the basis of their need satisfying attributes. These attributes are: (1) Relative advantages (2) Compatibility (3) Complexity (4) Divisibility (4) Observability and (6) Marketability.

i) Relative advantages: Women in rural areas are very much calculative in investment. They consider in what ways the technology will be helpful to them. Relative advantage is the degree to which a technology perceived as better than the idea it supersedes. It implies that every technology is supposed to substitute the earlier ones, otherwise why an individual would opt for it. Relative advantages are considered in three aspects. (i) Economic profitability (ii) Low initial cost or investment and (iii) Low perceived risk. In the scale of profitability and risk involvement the matrix would be like:

Table 4.4: Profitability and risk factors

Profitability	Risk involvement	
	High	Low
High	1.1	1.2
Low	1.3	1.4

Combinations

1.1 = High Profitability High risk= Large famers may consider

1.2 = High Profitability Low risk= Excellence for all farmer categories

1.3 = Low Profitability High risk= Rejected by all

1.4 = Low profitability Low Risk= Applicable for average farmers, i.e. traditional farmers

In this context we can argue in favor of traditional farming (sustainable agricultural practices) which has unique advantages like:

1. Low input agriculture
2. Low risk agriculture
3. Low use of energy/units
4. Very less soil problem
5. Decreased discomfort
6. Saving in time and effort
7. Eco-friendly
8. No residual harmful effect
9. Easy availability of inputs
10. Consistent with farming system

ii) Compatibility: Compatibility is the degree to which technology is perceived and consistent with the existing values, past experience and need of the adopters. Sustainable agricultural practices fulfill all these requirements like, social value, previous ideas and needs.

iii) Complexity: It is the degree to which the technology is perceived relatively difficult to understand and use. The sustainable agricultural practices are simple to adopt. These practices were undertaken by the farmers who did not have formal education. The sustainable practices could be examined in modern parameters and realized that these practices are simple to use.

Table 4.5: Scale to measure complexity of the sustainable agricultural practices

Si. No.	Practice	Simple to Complex				
		1	2	3	4	5
1.	Mixed farming					
2.	Crop rotation					
3.	Green manure					
4	Compost					
5.	Ecological pest control					
6.	Vermi compost					
7.	Integrated crop management					

The practices can be rated by the farm women in the scale of simple to complex to find out complexity of the technology. The rural women definitely feel that the practices as simple as have been practiced from generation to generation prior to green revolution.

iv) Triability of technology: Triability is the degree to which the technology may be experimented on a limited scale to create confidence in the minds of the adopters. In case of sustainable practices in agriculture we can apply the formula of Production input, Applicator and Motivational input (PAM). Production input is one which enters into production process like, seed, manure, fertilizer, *etc.* Applicator is one which helps production input to enter into the process of production like implements. Motivational input is one that adopter uses applicators on production inputs like attitude, skill and knowledge.

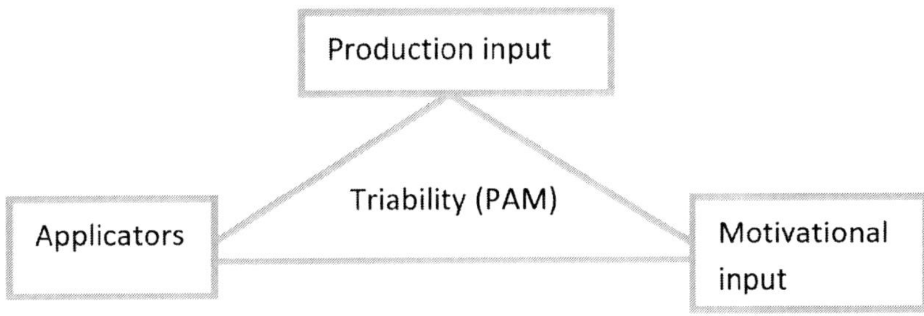

Fig. 4.7: Triability (PAM)

Box No. 4.2. Success of extension professional attributes to
1. Extent of contact
2. Commitment to bring change
3. Effectiveness in transfer of technology
4. Have empathy for client
5. Effective use of teaching aids

v) Observability: Observability is the degree to which the result or output is visible and can be communicated to others which can help in acceleration of rate of adoption. Both result and method demonstration is helpful to bring visibility in adoption of technology. The observability of technology can be assessed in the scale taking all the sustainable practices into the consideration.

Table 4.6: Scale to measure visibility of the sustainable agricultural practices

Sl.No.	Sustainable Practices	Range of observability/visibility				
		1	2	3	4	5
1.	Mixed farming					
2.	Crop rotation					
3.	Compost					
4.	Green manure					
5.	Vermi-compost					
6.	Ecological pest control					
7.	Integrated crop management					

vi) Marketability of produce: Farmer particularly women are more conscious about profit. Marketability is the degree to what extent demand exists in the market for the produce. Marketability includes selling cost, place of market, cost of production and profit. All these factors together decide whether a household is willing to opt for adoption of technology.

V. Extension Professionals and Adoption of Technology by Farm Women

An Extension Professional is an individual who influences farmer/farm women to adopt technology. He may be an employee of government, NGO, company, trading unit or business house. His interest is to influence technology he wants to propagate. The VAW, Lady VAW, extension officers, scientists when wish to farmers they act as extension professionals. They play very important role in adoption of technology because they make personal contact and influence for adoption. Their effective roles are:

1. Develop a need for the technology in farm women
2. Establish rapport with farm women for exchange of information
3. Help in diagnosing the problems
4. Create interest for technology
5. Translate interest into action
6. Prevent women from rejection and discontinuance of the technology
7. Maintain living relationship with clients for transfer of technology

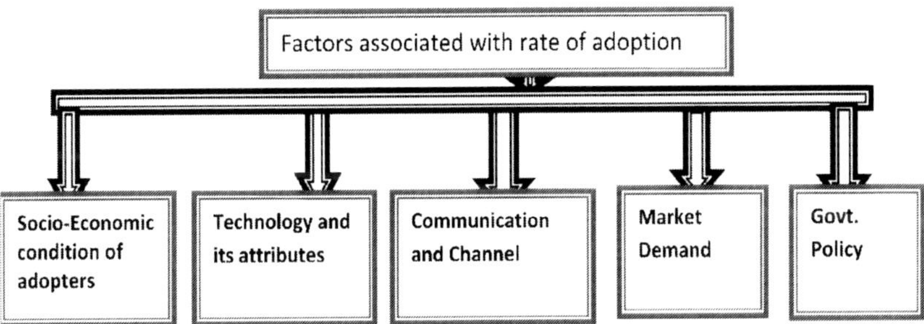

Fig. 4.8: Factors Associated with Rate of Adoption

VI. Socio-Economic Characteristics of Women Adopters

Socio-economic conditions regulate the activities of an individual. Women farmers relatively keep more vigilance on these aspects. A study conducted in the state to map the socio-economic variables in relation to level of adoption reveals interesting results. The level of adoption was studied classifying the adoption scores into high, medium and low level. The technology relating to crops like cereals, pulses, oil seeds and selected vegetables were taken into consideration. The technology relating to these crops were (i) seed and variety, (ii) seed treatment,(iii) fertilizer dose,(iv) plant protection measures, (v) cultural operation and (vi) post harvest care. The cumulative scores obtained on these accounts were taken to classify the farmers in the adopters groups of high, medium and low.

Data contained in table above reflect that in many cases agriculture produce are not enough to sell in market. The women are not much exposed to training programs to increase their knowledge, skill and attitudes.

Sensitizing women farmers: In gender studies there has been much debate as how to sensitize women farmers for better agriculture along with their social upgradation. To achieve this end, there is need to sensitize all concerns with women in general and agriculture in particular. The broad category audience who are to be sensitized are: (i) Public in general (ii) Grass root level extension workers (iii) Development personnel

Table 4.7: Socio-Economic variables and adoption level N=240

Sl.No.	Socio-Economic Variables	Level of adoption					
		High (43)	Medium (77)	Low (120)			
		f	%	f	%	f	%
1.	**Age group**						
	(i) Young	8	18.60	15	19.48	24	20.00
	(ii) Middle	13	30.23	30	29.87	36	30.00
	(iii) Old	22	51.17	39	50.65	60	50.00
2.	**Education**						
	(i) Illiterate	5	11.62	8	10.38	13	10.83
	(ii) Primary	12	27.90	21	27.27	32	26.66
	(iii) Middle school	15	34.88	26	33.76	41	34.16
	(iv) High school	8	18.60	14	18.18	22	18.33
	(v) Above high school	3	7.27	8	10.41	12	10.02
3.	**Family Composition (size)**						
	(i) up to 4 members	25	58.13	45	58.44	72	60.00
	(ii) 5 and above	18	41.87	32	41.56	48	40.00
4.	**Family type**						
	(i) Joint	10	23.25	16	20.00	20	16.66
	(ii) Nuclear	33	76.75	61	79.23	100	83.34

Contd.

Si.No.	Socio-Economic Variables	Level of adoption					
		High (43)		Medium (77)		Low (120)	
		f	%	f	%	f	%
5.	**Caste Composition**						
	(i) ST	7	16.27	14	18.18	18	15.00
	(ii) SC	8	18.60	12	15.58	20	16.66
	(iii) OBC	20	46.51	28	36.36	66	55.00
	(iv) Other caste	8	18.62	23	29.88	16	13.34
6.	**Membership in formal Organization**						
	(i) One	8	18.60	12	15.58	6	5.00
	(ii) Two	6	13.95	6	7.79	8	6.66
	(iii) Nil	29	67.45	59	76.63	106	88.34
7.	**Family Occupation**						
	(i) Farming	24	55.81	57	74.02	67	55.00
	(ii) Service	11	25.58	6	7.79	12	10.00
	(iii) Business	8	18.61	8	10.38	15	12.50
	(iv) Wage earning	0	0.00	4	5.19	22	18.33
	(iv) Any other	0	0.00	2	2.64	4	4.17
8.	**Land holding of the family**						
	(i) Up to 1 acre	3	6.97	36	46.75	60	50.00
	(ii) 1.1 to 2 acre	4	9.30	17	22.07	24	20.00
	(iii) 2.1 to 3acre	5	11.62	14	18.18	20	16.66

Contd.

Si.No.	Socio-Economic Variables	Level of adoption					
		High (43)		Medium (77)		Low (120)	
		f	%	f	%	f	%
	(iv) 3.1 to 4 acre	10	23.25	8	10.38	16	13.34
	(v) 4.1 to 5 acre	12	27.90	2	2.62	0	0.00
	(vi) More than 5 acres	9	20.96	0	0.00	0	0.00
9.	**Animal wealth of family**						
	(i) Up to no. 5	32	74.41	61	79.22	98	81.66
	(ii) More than 5 nos.	11	25.58	16	20.78	22	18.34
10.	**Cropping intensity**						
	(i) Mono crop	5	11.62	18	23.37	31	25.83
	(ii) Two crops	30	69.76	48	62.33	80	66.66
	(iii) Three crops	8	18.62	11	14.30	9	7.51
11.	**Use of electricity in farming**						
	(i) Yes	41	95.34	63	81.81	103	85.83
	(ii) No	2	4.66	14	18.19	17	14.17
12.	**Agriculture produce**						
	(i) Just manageable	8	18.60	38	49.35	18	12.50
	(ii) Less than family requirement	18	41.86	28	36.36	87	72.50
	(iii) Marketable surplus	17	39.64	14	14.29	15	15.00
13.	**Training Received on farming**						
	(i) Yes	8	18.60	12	15.58	21	17.50
	(ii) No	35	81.40	65	84.42	99	82.50

in charge of design and implementation of rural programs (iv) Scientists of agricultural domain (v) Planners and Programmers (vi) NGOs at village level (vii) Educational authorities and (viii) Women institutes dealing with problem of rural women.

There are certain principles that would help extension workers to achieve tangible results.

(i) **Reaching and Teaching rural women:** Gender sanitization requires determination and social responsibility. The grass root extension functionaries working in various developmental projects are required to reach and teach farm women about the production and productivity. This is not an easy job. Those who sincerely desire to bring a change in life style of women should find out extension techniques to reach them. Reaching farm women for sustainable agriculture the extension workers must know social values and culture of the target groups. It is more required in tribal areas where resistance to change is comparatively more. Knowledge about their language, their social system, liking and disliking would serve as important aids in reaching and teaching rural women.

(ii) **Determine the gap in knowledge, skill and attitude:** To reach rural women and work with them for specific objectives requires skill for determination of needs basing on the existing situation. There are different methods to calculate gap in need. What is present situation and to what extent the women have to move is the gap that we look for extension purpose. There is need to classify needs like (a) simple need and complex needs (b) individual need and community needs and (c) felt need or unfelt needs. The need analysis in systematic manner can be determined by applying PRA (Participatory Rural Appraisal) methods. When needs are integrated into the programs of development better tangible results are obtained as has been reported by Extension scientists in many cases. Gap comprehension in knowledge (Things known), skill (Things done) and attitude (Things felt) provide very good ideas to move ahead and achieve results in rural and tribal areas.

(iii) **Demonstration based extension approach**: To make sustainable agriculture popular, the adoption approach should be based on demonstration. As discussed earlier there are about seven to eight important practices in sustainable agriculture. These practices can be well demonstrated applying the principle, "learning by doing and seeing is believing". Extension functionaries or NGO change agents can very well adopt this approach with active participation of emerging SHGs.

(iv) **Group dynamics as the tool of implementation:** The upcoming SHGs will be of great help to start sustainable agriculture. Group philosophy, group structure, function, homogeneity and cohesiveness can be handled carefully to approach rural women to strengthen the organic farming ideas. Local extension worker should create a living relationship with client system so that technologies beamed through them are accepted and implemented. Creation of confidence in the minds of the target group is much more important as it brings permanent change. Extension worker, well tested technology, favorable situation and constant follow up with monitoring have brought success at many places. These experiences can be put into action in mobilizing rural women for sustainable agriculture.

(v) **One does not fit all**: Sensitization approach can be planned keeping in view the farm families of different farm sizes on which they are tagged as marginal, small, medium and large farmers groups. Small farm operation has specific problems. The farm development program should base on the formula of ***Situation + Objective +***

Problem + Solution. Analysis of each element provides required input to achieve the objectives in a systematic manner.

Women are not only the carrier of human race but civilization and sustainable development rests on them. They are the best upholders of environmental, ecological and social balances. Women comprise 50% of our population, contribute 75% work hours and receive 10% income and 1% share in property (FAO). They are invisible workers. Although knowledge is the base and human right of every individual as it creates consciousness and interest that leads to understanding, action and self confidence yet facts reveal wide gender gaps. They are discriminated on education, training that are essential and useful weapons in the competitive world of productivity, production, equity, sustainability and empowerment.

Everywhere, the starting point is to empower women to ensure their participation in decision making process that affect their lives and enable them to build their strengths and assets. Building the assets of the poor and empowering them is the starting point of eradicating poverty. The strategy for women empowerment needs,

(i) Policy reforms and actions that would enable to have access to resources

(ii) To ensure education and health to all

(iii) Social safety

The implications of the above need based strategy are:

(i) Removal of all discrimination against girls starting from birth in all aspects of survival, health, education and upbringing.

(ii) Empowering women by giving them equal rights and access to land, credit and job opportunities.

(iii) Taking action for ending all forms of violence and abuse against women.

Empowerment is a process not a thing which cannot be given as such. The process of empowerment is both individual and collective. The women empowerment can be considered as continuum of several inter-related, interwoven and interactive mutual reinforcing components.

These are

1. Awareness building about situation, position, discrimination, rights and opportunities, group identity and wider recognition.
2. Capacity building about planning, decision making, organizing, managing and implementing activities to deal with people and institutions.
3. Participation and control about affairs in the home, community and society.
4. Action about gender equity between men and women for being recognized and respected as citizens and human beings with a contribution to make at all levels of society not only in home.

CHAPTER 5

Gender Based Participatory Extension Approach for Sustainable Agriculture

Teaching, Research and Extension are interrelated and interwoven in development. It may be in any field, like agriculture, animal husbandry, fisheries, forestry etc. Whenever we talk of extension, it implies the participation of clients in the program. Extension can never be defined without inclusion of participation of people. Extension approach means application of tools and methods by which we can reach people. The participation is the core value of extension but its use differs from situation to situation bringing out a degree of differences leading to variation in result. In extension the farmers are advised and the blue print prepared for them are recommended to be adopted. Such approach did not yield desirable result, as result regarding realization was under question the impact of extension on farming communities. The studies conducted on extension approaches tend to reveal poor result as the participants failed to own and shape the program designed for them. We call it passive participation as the program was designed and implemented by the scientists/ extension workers specifying the role of farmers as land giver and labor given along with receiver of the output. Such approach did not help to bring sustainability and the practices ended with end of external funding may of kind or cash. Gradually attention was paid to understand the meaning and concept of participation and the attempt tended to real participatory one.

Government and Non-Government Organizations are recognizing the need to move away from instructions and blue print solutions towards more participatory approaches to support communities in development. The principle has to accommodate the concept that people are the owners and shapers of the development. For Agriculture extension workers, the challenge is to change their ways of contact and learning through participatory approach.

I. Extension Professional as Facilitator

With realization of failure, a move was made to train extension professional as facilitator in many of the under developed and developing countries and evidences are found to be encouraging. The concept changed from extension agent to facilitator emphasizing learning to interact closely with social groups and communities, better listeners and facilitators and developing a responsive two -way communication process between the community and local technology institutions. The principle of extension discipline, "working with and through the people and not for the people" can make extension participatory. There is need to understand the inner meaning of extension at field level to illicit true participation of clients.

Extension Professionals have now preferred PRA (Participatory Rural Appraisal) as one of the good tools to initiate extension activities in development sector. PRA provides good illustration, nice diagram and different maps of different beautiful colors but their inclusion in program development is not reflected. Therefore to achieve participatory extension, the extension professionals are to be trained in the methodologies of Participatory Extension Approach (PEA).It is important to be trained in learning process rather than in teaching process and methodologies. The task of training thousands of extension workers in PEA requires clear understanding of concept behind PEA. If the field level extension workers are to be able to implement PEA successfully they require practical skill and methods. The more important is in-depth understanding of learning process. A systematic approach has to be evolved to harmonize different approaches advocated by experts. The experiences gained in different countries in implementing PEA needs analysis to suggest for adoption.

Extension worker ought to have managerial skills and these skills should appropriately reflect in capacity building of the target groups. A competent farmer should have these skills to make the profession sustainable. These are (i) Conceptual skill (ii) Human skill and (iii) Technical skill. The conceptual skill is the ability of an individual to understand how the components of technology are interdependent. In case of farming, the components like plant nutrition, irrigation. Plant protection is interdependent. The question is how to maintain balance in operation of these components so that optimum output is achieved. These abilities of both extension workers and farmers are equally important to achieve participatory extension approach. The human skill involves working with people. The farmers have to work with farm labor, neighbours, extension workers, scientists and traders in marketing. The abilities to resolve conflicts motivate people to work around and cooperate and good communication are required to compete at present. Technical skill is the ability to make use of technology in appropriate manner. It involves knowledge, expertise and working procedure. All these skills are now required for farmer or farm women to sustain their professions.

II. Characteristics of PEA

It we accept the PEA as best method to evoke positive response of the clients; we have to understand learning process. The PEA based on participatory concept has the following characteristics.

- **Community mobilization:** PEA requires integration of community mobilization, planning, extension approach and research into one component so as to develop a community -need -based program.
- **Equal Participation:** PEA believes equal participation of farmers, extension workers and researchers who can learn from each other and contribute their knowledge to development.
- **Management ability:** The focus of PEA is to enhance managerial and problem-solving ability of people for which increased emphasis should be reflected in planning process.
- **Innovativeness of client:** Promotion of innovativeness of clients in refinement/modification/development of location based technologies is the greatest strength of PEA which can bring permanent development in the concerned area. This implies to capacity building of the farmers/farm women through appropriate training program.
- **Action learning:** PEA believes action learning of the farmers in the line of "Learning by Doing". Encouraging target groups to learn through experiment, building their own knowledge and practice and blending them with new ideas is the real extension approach leading to successful participation. According to Kingsley and Garry (1957) "Learning is the process by which behavior in the broader sense is originated or changed through practice or training".

- **Heterogeneity not homogeneity:** Rural households cannot be scaled in homogeneity for extension program. PEA recognizes difference in respect of social groups, conflicts and differences, interest, knowledge level, skill, ability of client groups which serve as pillars of success. Small, marginal, medium and large landholders are different in many aspects for which the principle of Extension, "there cannot be one program for all" has applicability in all rural development planning, *Extension is like a school of trying, where you try out ideas and share your experience with others"*.

Our confusion has been due to use of variety of words like, approach, method, tools to explain what is participatory. To make it clear, concept provides frame work, frame work provides strategy, strategy provides methodology and methodology provides tools to work.

Concept → Frame work → Strategy → Methodology → Tools

PEA as understood in extension approach involves transformation in the way extension professional interact with farmers. Joint learning in community is the central to PEA. There is great difference between PEA and PRA. PRA is based on useful tools for participatory analysis and interaction with rural people. PRA is tool box not extension approach and PEA is the vehicle to carry PRA. In extension, selection of these methodologies and tools would include adult learning, group extension methods and farmer's field schools, extension program planning etc along with PRA or RRA.

III. Development as Learning Process

There has been constant effort to increase impact of research and extension in rural development with special reference to agriculture. In all these efforts participation became central point.

To experiment the means and ways to secure farmer participation, different programs have been formulated reflecting the issues to learning and participation in development. These are, (i) On-farm research (ii) Farming system research (iii) Technology refinement (iv) Farmer to Farmer approach (v) National Demonstration (vi) Technology adaptation and many others in different countries. The basic aims behind these approaches were, (i) Outsiders are unable to determine best methods for rural people (ii) Building up of farmers management skill through practical field experiment (iii) Spreading of innovation through interaction of rural people (iv) Changing role of extension worker to be facilitators and (v) Farmer need based research.

IV. Role of Extension Professionals

i. Teacher to facilitator: Facilitation means providing the methodology in terms of communication, information flow, technical support and possible options.

ii. Coordinators: Extension worker to act as coordinator between farmers and knowledge centers/institutions

iii. Training: Training of community own facilitators/ leaders are to be trained by the extension workers.

iv. Document Recorder: The Extension worker helps in documenting farmers knowledge and experience and produces fact sheets for the use of the farmers.

We sometimes get confused about transfer of technology and Participatory Extension Approach. The transfer of technology is characterized by infusion of technology by outsiders, fixed menu, farmer to farmer extension mode and role of extension worker as teacher. On other hand, PEA is characterized by empowering of client, farmers are facilitated by outsiders, basket full of choices, selection of technology according to choice of farmers and farmer to farmer the transfer of technology. However, PEA all times acts as facilitators rather than teacher to the farmers.

Shifting Focus: In PEA the shifting focus is teaching to learning, from hierarchical, top-down to participatory bottom up approach, from centralized to decentralized decision making that enable/put institution under pressure for change as well.

PEA in Operation: The PEA in Extension has four major phases.

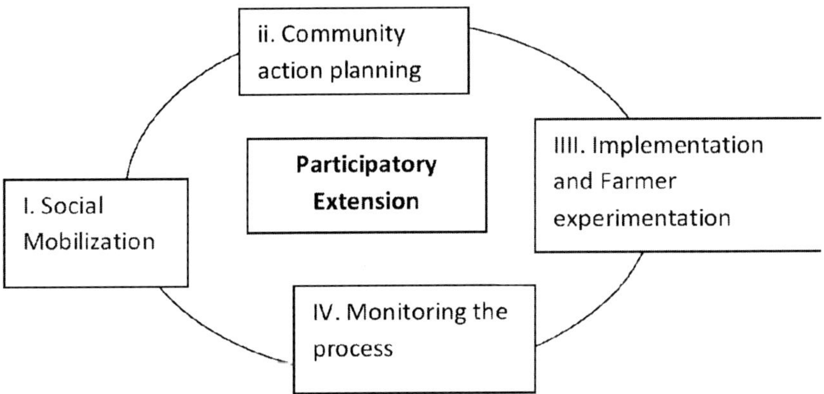

Fig. 5.1: Participatory Extension

V. Stages of implementation of PEA

Phase I Social Mobilization

It means facilitating community people to analyze their own situation. Social mobilization requires true and continued enthusiasm of the targeted population. This must by backed by community organization which can support process to take forward. In each location or situation there are institutions which support the process of development. These institutions are to be correctly identified by the program participants. The important point is that all institutions can not contribute because of diversity. Such institutions which can aid to process of development are to be identified and their capacity building is to be taken up. At this phase the important steps are,

1. Entering the community and building the trust
2. Identifying and supporting effective organizations
3. Feed back to community
4. Raising awareness in the whole community
5. Identifying community needs
6. Understanding difference in wealth
7. Understanding needs

1. **Entering the community and building trust**

Extension workers entering into a community to initiate development activities have to learn certain steps for success. These are,

(a) Arranging informal meeting with local leaders and influential people.

(b) Meeting local institutions and seeking their support

(c) Learning about perception of local people and institution

(d) Ascertaining problems and needs of community

(e) Building relationship and trust with local people

2. **Identifying and supporting effective organizations:**

(a) Enlisting local institutions and support areas

(b) Determining their strength, weakness and functions

(c) Understanding complete and alliance network of institutions

(d) Identify human and material resources

3. **Feed back to the community:**

The Extension worker after identifying supporting local institutions is required to give feed back to community to get their participation.

(a) Results are fed to the community

(b) Feedback should be of informal type

(c) Extension worker should be neutral in expression of opinion

(d) Initial selection of partner institution would help to move ahead

4. **Raising awareness among core members of the community**

The extension worker by now knows who are the people matter in the community and can establish rapport for participation.

(a) Organizing meeting with local leaders

(b) Care to be taken to select poor and marginalized households

(c) Initiation of action learning process

(d) Make people realize about need of the program

(e) Encourage people to analyze their problems and causes

(f) Initiate planning process

(g) Provide training for transformation

(h) Arrange awareness workshop

(i) Seek contribution of all to prepare a common platform for development

5. **Identify community needs**

Need identification is a skillful and knowledge oriented process. The extension worker must have experience and training as how to identify community needs.

(a) Let community decide the needs of their community

(b) Understand difference in status, wealth and perception of people

(c) Use need assessment techniques

(d) Identify what people indicate as poverty and causes

(e) Understand perception of different people, their priority and needs

6. **Understanding of difference in wealth**

(a) Initiate wealth ranking matrix

(b) Initiate need assessment

(c) Decide priority of need for each household

7. Understanding needs

(a) Discussion on different wealth ranks

(b) Involvement of members from each group

(c) Special care for poorest and marginalized households

(d) Include at least 10% of the total members of each group

Phase II Community Level Action Planning

Community level Action Planning by the community member is key to PEA. PEA emphasizes extension program as by the people, through the people and for the people. A good planning brings good result. In any community based action the planning should be as far as perfect to indicate action components. Planning means looking ahead. It is deciding in advance what is to be done. Planning answers 5ws. These are what, where, when, who and why. Decision making is very important element in planning. It is an intellectual process. Good planning helps in (i) management of objectives (ii) to offset uncertainty in operation (iii) to secure economy in operation (iv) to help in coordination (v) to make control effective and (vi) to increase organizational effectiveness. In community level planning PEA suggests the following steps.

1. Feedback to all members of the community

(a) Enable community to prioritize the needs

(b) Find out all causes of problems and possible solutions

(c) Identify local institution which can be of good use

(d) Draw schedule of work on identified needs

(e) Find out criteria or indicators for measurement of progress

(f) Collective decision making and ownership of the project is the aim of PEA

2. **Prioritization of problems and needs**

(a) Formation of small groups according to gender, age and institutional membership

(b) Collect feedbacks and put in planning process

(c) Common discussion and negotiation to arrive at decision

(d) More than one problem can be listed for action

(e) Find clear-cut idea about root causes of problem

3. **Searching for solutions**

(a) List out possible solutions

(b) Common discussion on the possible solutions

(c) Pooling ideas from all members

(d) Keep scope for generation of new ideas at later stage

(e) Decision on how solution can be tested, who will coordinate with what responsibility

(f) Emphasis on decision of ITKs

4. **Look and learn tours**

(a) Visit success project

(b) Right selection of members to visit or go on exposure visit

(c) Plan such visit ahead with all information

5. **Mandating local institutions**

(a) List of local institution to extend cooperation

(b) Decide the responsibility of institutions

(c) Choosing leaders to clarify goals, communication for all purposes and implementation

(d) Decide accountability and commitment of institutions

6. **Action planning**

(a) It is time to develop plan of action to be developed by community members

(b) Define the role and function of extension worker as facilitators

(c) Better to start on small scale.

7. **Priority area and indicator for measurement of Progress**

(a) Community to decide as how to measure progress

(b) Decision on indicators of measuring progress

Phase III. Implementation and Farmer Experimentation

Implementation starts with final decision about the procedure and solutions. Once the procedure is decided and solutions to the problems are settled, the implementation starts. It needs to decide the plan of work, calendar of operations and procedures for evaluation. The program of such nature indicates (i) situation (ii) objective (iii) problem and (iv) solution. These four components basically regulate the implementation part in PEA. Rural Development Programs which people decide with efforts of extension workers clearly set the following points to be kept in mind.

(a) Learning to experiment the ideas

(b) Encourage farmers to experiment their idea

(c) Combine new with old ideas and make synthesis

(d) Create confidence in the mind of the farmers

(e) Provide opportunity to select best option

(f) Invite resource people and learn from them

(g) Simple paired design will make better understanding of the result

(h) Exchange experience with farmers in informal way

(I) Learning through experiment is the best of all

(a) **Learning through experiment the ideas**: The implementation phase of PEA is also called farmer experimentation. Here learning process is the important core issue. At this phase the extension worker is to encourage farmers to experiment with ideas and techniques emanating from farmer's knowledge or from outside source. It is yet to be confirmed that whether farmer knowledge or outside knowledge which one performs well in giving additional benefits to the clients. An interaction between and among the knowledge sources is required to arrive at conclusion.

(b) **Encourage farmers to experiment their idea:** Idea and its translation to action are crucial to PEA. The PEA strongly argues for use of farmer's knowledge.ITK or people's wisdom which sometimes has no scientific basis. During implementation and experimentation process, new questions and problems not seen at the beginning may arise and become action research agenda. The situation will provide good scope for research scientists in solving farm problems. This aspect is lacking in many cases. Our attempt is to take solutions to the community and hardly think of collecting researchable agenda from them.

(c) **Combine new with old ideas and make synthesis:** Agriculture in India from ancient time had undergone various metamorphoses. The change in farming practices happen as per change in climate and change in choice of the farmers. Sustainable agriculture has many good practices. When we are armed with new technologies, it is better to compare old and new and find out an adoptable practice suiting to local condition. The ITK (Indigenous Technical Knowledge) are environment friendly, easy to use and cost effective. The attempt needs a thorough understanding of science behind the practice. Many farmers do this job at their level like refinement of technology in application of fertilizer, sowing time, irrigation and plant protection.

(d) **Create confidence in the mind of the farmers:** Creating of confidence in the mind of the farmers or farm women is not easy job. Creation of confidence on practice comes later and on extension worker first. Confidence building in farming communities follows the principle" seeing is believing". Households choose the options and ideas they think are most responsive to their problems. Conducting simple comparisons between conventional and new technique can be powerful learning leading to confidence building. In extension we use both method and result demonstration to create confidence about the practice. Before that conduct of the extension workers and their knowledge level also adds to confidence building.

(e) **Provide opportunity to select best option**: The farmers and farm women faced with new technology undergo a series of decision making steps before drawing a final decision. The factors like (i) relative advantage (ii) simplicity in adoption (iii) compatibility to farming system and (iv) affordability are considered before adopting a practice. Applying these criteria the option should be exercised. In farming operations, the decision for new technology comes from family as a whole. It is therefore advisable to involve both male and female members to arrive a solid decision.

(f) **Invite resource people and learn from them:** The principle is, more the exposure better is the learning process. The PEA is based on learning concept. The learning concept is change may be positive or negative. The role of extension worker is for positive response. Extension worker should acquire the knowledge of learning process. The factors that determine learning are (i) motive (ii) stimuli (iii) generalization (iv) discrimination (v) response (vi) reinforcement and (vii) retention. The experts, scientists, experienced farmers and farm women would accelerate the process of learning and the target group will derive benefits of it.

(g) **Simple paired design will make better understanding of the result:** Simple paired design creates interest and convincing to the participants. Better to take one practice and compare with local practice to find out difference. This is the simplest design with high turnout. Further the requirements and preparation are also less. Under village condition paired design is mostly preferred. Farmers and farm women are to observe, compare and analyze themselves. It helps to understand factors which contribute to differences

(h) **Exchange experience with farmer's informal way:** The greatest benefit is achieved when participants interact among themselves. The process brings positive, negative, deficiency and mistakes occurred during implementation of the program. The individual clarifies the doubts and get enlightened. More interaction better is the clarification of doubts. When the concept is examined in the light of JOHHARI Window the dark area get reduced with increase of interaction on the subject.

I Know	I don't know
Public areas (Open self)	Blind area (Blind self)
Private area (Open self)	Dark Area (Undiscovered self)

(I) **Learning through experiment is the best of all:** Experiment done by individuals provide permanent learning. Own experiment is the best teacher in life. Same principle is applied in case of farming communities. Those who take up practice and complete may be success or failure, know all plus and minus points which strengthen their confidence.

Phase IV Monitoring and Evaluation

It may be of (a) mid –season or (b) total process monitoring with emphasis of visiting project area, sharing with experience and confidence building measures. Mid season evaluation of the experiments and new techniques has special importance. In middle of the crop season before crop is harvested, farmers/ farm women with help of extension workers organize an evaluation of the field performance of the different ideas and techniques they have tried. In this case the innovativeness of the idea should be central theme in which number of trials, management of field and quality of presentation should play vital role. The major objective of evaluation is (i) sharing of knowledge (ii) building of confidence and (iii) encouraging farmers for more adoption of the technologies. The community members are to arrange a review meeting sufficiently ahead of the season. The meeting should review the process, evaluating planned activities and the indicators for success during planning phase. The meeting should consider the issues like leadership, strengthening of capacity of the unit and community participation with special consideration of inclusion of poor and marginalized households.

Besides, PEA emphasizes process review, self evaluation and planning for next season. It helps in (a) location of leaders (b) strengthening of institution capacity (c) community participation and (d) inclusion of poor and marginal households.

Possible Problems of PEA: In operation, PEA will face a number of problems. These need experience and expertise to overcome. The important possible problems are:

(a) Dominance of leaders

(b) Dominance of men over women

(c) Dominance of progressive farmers

(d) Conflicts in community

(e) Problems in need assessment

(f) Problems of funding

(a) **Dominance of Leaders:** In real life situations, the leaders or educated people in group make all decisions and others simply make it O.K. and many times remain silent giving nod to the proposal. In watershed programs, the village leaders make all decisions. In this case the poor and marginalized families specifically women farmers fail to express their opinions. Because of illiteracy and poverty many members refrain from taking active participation. In case of water shed, grain bank, child welfare and awareness campaign such situations are observed. To make PEA effective the steps like (i) organizing of training (ii) convincing leaders personally (iii) influencing institutions stressing on role and responsibility and (iv) encouraging individual members to express their ideas freely would help to overcome the problem.

(b) **Dominance of men over women:** Ours is a male dominated society. Although all state governments are enacting laws for participation of women in all fronts still women are not coming forward. The reason is of social system. In presence of elder men, the women prefer to remain shy rather than expressing their views. Our observation reveals that in community meetings hardly 3-5% of the women members come forward to express their views. This is more or less in tribal areas.PEA in such cases is not allowed to operate in true spirit. The reports indicate that in formally elected members at Panchayat organization are controlled by men in their actions. Again political system also imposes some restrictions in political party matters. The solution of this problem has to be found out at local level. The steps like (i) inviting men and women constantly to meetings (ii) involving both men and women as members (iii) raising gender issues and (iv) creating confidence in their mind would help to move forward. At present SHGs are coming up empowering women for active participation. The SHGs can be taken in development issues for better results providing greater opportunity to women to play their due roles.

(c) **Dominance of progressive farmers:** Under T&V system of extension approach, it is our experience that progressive farmers in community take lead part and avail maximum benefits leaving fellow farmers behind. More attention of development authorities, frequent contact at different hierarchy make progressive farmers to create a class or status which keep them away from fellow farmers. The suggested steps to overcome the problems are (i) introduction of inter community competition (ii) making progressive farmers understood about their roles and expectation and (iii) guiding them at important interaction sessions.

(d) **Conflicts in community**: In real sense due to frequent election system and increasing influence of political parties there is conflict in all communities. It has become so that in each and every developmental activities, the views of political parties are reflected as a result many projects are faced with resistance. Community conflicts are of many kinds. The conflicts relating to livelihood system or earning of money takes serious proportion if not solved at the beginning. The conflict management aspect has direct bearing on implementation of PEA.

(e) **Problems in need assessment:** Need assessment and project initiation sometimes become visible where some are excluded from the development. Normally the problem realized by the majority draws attention of authorities. As per principle there cannot be one program for all. Taking status, gender, age, poverty and other factors the community can decide program for all. The extension worker applying principle of program planning can act as facilitator to develop need based conflict free programs for development.

(f) **Problems of funding:** For development funds are provided by state, national or international organizations. Some donors of national and international base also provide funds. Observation and experience reveal that sustainability of development program is fund based. Once fund flow is stopped the programs are stopped. Attempt for generation of resources and their balance use be made to overcome the problem of funding.

Problems observed in Extension Work are almost same in case of PEA. In Extension field operation problems are encountered in different aspects. While dealing with farm women in PEA the problems are found in case of (i) Economic system (ii) Socio-cultural system (iii) Geo-physical situational system (iv) Technology transfer system and (v) Follow up system. The overall aim of extension is to bring farming communities to a better status in respect of all walks of activities. Since adoption of technology occurs in a social system the social problems become prominent that too more in case of women. The resource poor and marginalized families become victims of it.

To overcome the problem and push forwards scientific approach in existing farming communities the PEA is becoming more acceptable as it operates in ownership of the farm families. The steps to be followed in PEA are illustrated.

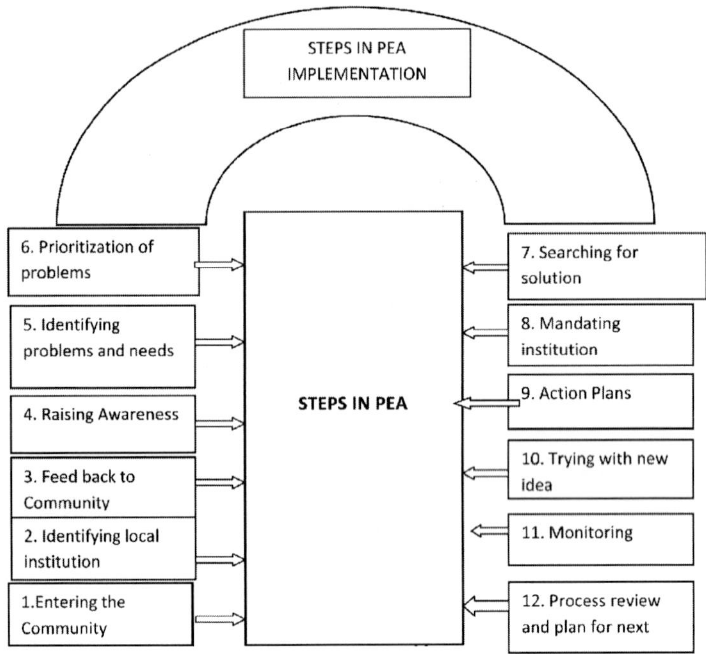

Fig. 5.2: Steps in PEA

Criteria for Successful PEA

1. Client participation and involvement
2. Empowerment in terms of increased articulation confidence and decision making
3. Implementation of community projects
4. Experimentation with ideas
5. Documentation of experience, learning and knowledge

For measurement of the progress or evaluation, it is desirable that participants should identify indicators. The indicators so selected would reveal the strength and weakness of the projects and impact of PEA. This would provide a frame work for PEA. Impact of PEA is to be determined at the level of stakeholders, farmers, farm women, institution, community and extension worker with special reference to adoption of technology suggested. Farmer's problem solving, management capacities and other qualitative indicators can be used to evaluate effectiveness of PEA.

Approach to make the PEA process transparent and operational, a detailed conceptual framework and description is required. The steps suggested should not be taken as blue print. There may be scope for addition or deletion of steps. A high quality training and coaching in the field as well as peer learning groups needs to be established to achieve quality implementation. Scope of PEA approach to social mobilization and local institutions has potential beyond agriculture. Issues like improved self governance and decentralization, better representation and accountability for local leaders are indicated as PEA has more to offer than just agriculture technology.

Implementation of PEA is subject to interest and commitment of extension worker who are accountable from top to bottom approach. Institutional reform, change in existing system and resource implication are individual important factors in total effect. The major difference with PEA for which it emphasizes is to change attitude and behavior of extension agent. However, equal impact may not be expected at all cases since it is location specific in approach.

Motivational incentives for Participatory Approach in Extension

Motivation or stimulation for agro business is relatively a difficult job. Still the date, agriculture and agri products are not attached with business value as in case of industrial business. Once we feel to provide employment to rural households who are unemployed and in search of livelihood would join PEA for agriculture. After globalization, earning money to live is not very difficult. Therefore selecting right type of agribusiness requires deep analysis. Our attention has been to attract farmers for profitable farming. But to make farming sustainable, we have to select and train farmers for active participation.

Motivation as applied in the case of agriculture, it comes from two sources. One internal and other is external sources.

1. **Internal Motivational factors**

- Desire to be in agribusiness
- Desire to take own decision for agri-business
- Occupational and educational back ground
- Training availed if any

2. **External Factors**

- Government assistance. Such assistances are available
- Availability of financial support by credit agencies at lower rate of interest.
- Market demand
- Technical/business training
- Support from parents/relatives
- Sympathy of Government

The above forces can act only if there is entrepreneurial behavior with the farmers. These are,

1. Need for achievement
2. Creativeness and innovative in nature
3. Strong desire to stand on own strength
4. Ability to advocate for Agro-Entrepreneurs
5. Believes in change and accepts change

References

Altieri, M.A. (1999). Biodiversity and Pest Management in Agro-Eco-System. Haworth Press, New York

Desai, B.K. and Pujari B.T. (2007). Sustainable Agriculture-A vision for Future, New India Publishing Agency, New Delhi

Gulati, J.M.L. and Barik,T (2011). Orissa University of Agriculture and Technology, Bhubaneswar, Odisha

Hogmann, J. Chuma, E. Muriira, K. Connolly, M. (1999). Putting Process into practice-Operationalizing of Participatory Extension, Agriculture Research and Extension Network, Network Paper 94, Odi.

Kingsley, H. L. and Garry, R. (1957). The Nature and Conditions of Learning (2nd ed.), Englewood Cliffs, New Jersey : Prentice - Hall.

Milkah, K, and Reddy, P.R. (2007). Gender Issues, Haritha Publishing House, Secundarbad 50010

Pezzey J. (1992). Sustainable Development Concept: An Economic Analysis. Washington, D.C.,

Satapathy C and Mishra, S. (2007). Women at Work, Golden Graphic Bhubaneswar

Satapathy, C and Mishra, S (2008). Extension Techniques for Rural Management, Kalyani Publisher, New Delhi

Sharma, A.K. (2008). A Hand Book of Organic Farming Agrobios, India

Soul, J.D. and J.K. Piper (1972). Farming in Nature's Image: An Ecological Approach to Agriculture. Island Press, Washington, D.C.

World Bank, (2012). Gender equality and Development, World Bank Washington, D.C

Terminology

A

Abiotic factor: A non living component of the environment, such as soil, nutrients, light, fire or moisture.

Acid soil: A soil with a pH value <7.0 usually applied to surface layer or root zone.

Adaptation: (i) Any aspects of an organism or its part that is of value in allowing the organism to withstand the condition of the environment. (ii) The evolutionary process by which a species genome and phonotypical characteristics change over time in response to change in the environment.

Adaptive research: It aims at problem solving. It takes existing technology and tailors it to define areas for a definite group of farmers who need solution to their problems.

Adoptive research: Adoptive research aims at developing new inputs, new varieties, mechanize, packages, etc.

It uses theory and tools for basic research to develop new technology.

Agricultural wastes: Agricultural waste materials can be put to use because for improvement of soil health, soil fertility, physical condition of soil, it acts as source of energy.

Agriculture: Agriculture is that sector of activity in which there is greater interaction between environment and human culture, which has grown in and from it.

Agri-silviculture: The conscious and deliberate use of land for the concurrent production of agricultural crops and forest crops.

Agro biodiversity: "is a fundamental feature of farming systems around the world. It encompasses many types of biological resources tied to agriculture, including: genetic resources-the essential living materials of plants and animals; edible plants and crops, including traditional varieties, cultivars, hybrids, and other genetic material developed by breeders; and livestock (small and large, lineal breeds or thoroughbreds) and freshwater soil organisms vital to soil fertility, structure, quality, and soil health; naturally occurring insects, bacteria, and fungi that control insect pests and diseases of domesticated plants and animals; agro ecosystem components and types(Poly–cultural / monoculture, Small/Large scale, rain fed /integrated etc) indispensable for nutrients cycling, stability and productivity.

Agro Ecosystem: Agro ecosystem are ecological systems modified by human beings to provide food, fibre or other agricultural products.

Agro-Ecological Zone: It is an area of similar soil, vegetation and population density characteristics resulting in similar types of cropping system.

Agro-forestry: Agro forestry has been defined as sustainable land management system which increases the yield of land, combines the production of crops, including tree crops and forest plants or animal simultaneously or sequentially on the same unit of land and applies management practices that are compatible with cultural practices of the local population.

B

Basic research: It aims at exploring the frontiers of science. It develops new theory and research tools.

Beneficial insects: Beneficial insects are predators, parasites or competitors of insect, pests, helping to regulate pest population without harm of crops.,

Bhoodan: it is a movement where it is suggested that farmers with surplus land may voluntarily donate land to landless.

Biodiversity: At its simplest level, biodiversity is the sum total of all the plants, animals, fungi and microorganisms in the world, or in a particular area; all of their individual variation; and all the interactions between them.

Biodynamic agriculture /Bio dynamic farming: Both a concept and a practice,biodynamics "owes its origin to the spiritual insights and perception of Dr. Rudolf steiner an Australian philosopher and scientist who lived at the turn of the century". Dr. Steiner emphasized many of the forces within living nature, identifying many of the factors and describing specific practices and preparation that enable the farmers or gardener to work in concert with these parameter. Central to the biodynamic method ... are certain herbal preparation that guide the decomposition processes in manures and compost.

Biological Control/Bio-control: "Biological control is, generally, man's use of a specially chosen living organism to control a particular pest. This chosen organism might be a predator, parasite, or disease which will attack the harmful insect. It is a form of manipulating nature to increase a desired effect. A complete Biological Control program may range from choosing a pesticide which will be least harmful to beneficial insects, to raising and releasing one insect to have it attack another, almost like a 'living insecticide.'

Biological farming/ Ecological farming: Biological and ecological farming are terms commonly used in Europe and developing countries. Although sometimes strictly defined ,e.g "Biological farming is a system of crop production in which the producers tries to minimize the use of "chemicals" for control of crop pests , both biological farming and ecological farming are term used in the broader sense, encompassing various and more specific practices and technique of farming sustainability, e.g organic, biodynamic, holistic, natural.

Biomass: The mass of all the organic matter in a given system at a given point in time.

Bio-resources Flow Diagram: It is a picture of natural resource type drawn as topographical cross sections of land and water resources.

Biotic factor: An aspect of the environment related to organisms or their interactions.

Bonded labour: It indicates a relationship which is unfree or tied involving two parties of unequal status. It is usually applied to relationship where a labourer or debtor borrows money from the creditors under a contract of work until the debt is repaid.

Buffer zone: A less-intensively-managed and less-disturbed area at the margins of an agro ecosystem that protects the adjacent natural system from the potential negative impacts of agricultural activities and management.

Bulk density: The mass of soil per unit of volume.

C

C:N ratio: The ratio of carbon to nitrogen in a material. The decomposition of materials is regulated in part by this ratio, so materials with different **C:N ratios** are usually mixed to improve decomposition rates in composting. The **C:N ratio** for optimal biological activity is about 25:1, with higher values being nitrogen limited and lower values being carbon limited. The average **C:N ratio** for soils is about 10:1.

Capillary water The water that fills the microporesol the soil and is held to soil particles with a force between 0.3 and 31 bars of suction. Much of this water (that portion held to particles with less than 15 bars of suction) is readily available to plant roots.

Carbon dioxide compensation point: The concentration of carbon dioxide in a plant's chloroplasts below which the amount of photosynthetic produced fails to compensate for the amount of photosynthetic used in respiration.

Carbon fixation: The part of the photosynthetic process in which carbon atoms are extracted from atmospheric carbon dioxide and used to make simple organic compounds that eventually become glucose.

Compensating factor: A factor of the environment that overcomes, eliminates, or modifies the impact of another factor

Compost: Mixed decayed and decaying organic matter with available nutrients useful for fertilizer.

Crop ecology: It is the study of inter-relationship between crop plants and the environment.

Cropping pattern: The yearly sequence and spatial arrangement of crops and fallow in a given area is known as cropping pattern.

Cropping system: Cropping system is a land use unit comprising soil, crop, weed, pathogens and insect subsystems that transforms solar energy water nutrients, labour and other inputs into food, feed, fuel and fibre.

D

Decomposer: A fungal or bacterial organism that obtains its nutrients and food energy by breaking down dead organic and fecal matter and absorbing some of its nutrient content.

Decomposition: The process by which materials are broken down into simpler compounds by decomposers.

Demonstration: Demonstration is a presentation that shows how to perform an act or to use.

Dew point: The temperature at which relative humidity reaches 100% and water vapor is able to condense into water droplets. The dew point varies depending on the absolute water vapor content of the air.

Diffusion: It is the process by which innovation spreads. The diffusion process requires ingredients like innovation, communication, over time and a social system.

Direct Matrix Ranking: DMR refers to arranging a group of items or things in order of importance. Ranking means placing or grading a thing according to the quality or achievements where as scoring refers to be certain number of points to a thing in a competition. Direct Matrix Ranking for technology decision behavior refers to placing different technologies in order of importance.

Diversity: (1) The number or variety of species in a location, community, ecosystem, or agro-ecosystem. (2) The degree of heterogeneity of the biotic components of an ecosystem or agro-ecosystem.

Dynamic equilibrium: A condition characterized by an overall balance in the processes of change in an ecosystem made possible by the system's resiliency and resulting in relative stability of structure and function despite constant change and small scale distrubances.

E

Ecological diversity: The degree of heterogeneity of an ecosystem's or agro-ecosystem's species makeup, genetic potential, vertical spatial structure, horizontal spatial structure, trophic structure, ecological functioning, and change over time.

Ecological entity: Distribution of land areas to create land based unit on the basis of symbiotic relationship form ecological entity.

Ecology: It is generally defined as the study of plants and animals in reciprocal relationship with their environment or external world.

Ecosystem: Any collection of organisms that interact or have potential to interact along with the physical environment in which they live form an ecological system or ecosystem. A functional system of complementary relations between living organisms and their environment within a certain physical area.

Experimental group: The group under study, which is subjected to manipulation and control. Greater control is exercised over conditions of observations to effectively eliminate the possibility of extraneous factors.

Exploratory studies: Exploratory study is undertaken when relatively little is known about the phenomenon to be researched. Its aim is to collect facts, which will later help to formulate a precise research problem or generate sound hypothesis for further research.

Ex-post Facto research: Ex-Post facto research design is defined as that in which independent variables have already occurred and in which researcher starts with observation of a dependent variable.

F

Farm Clinic: It is a facility developed and extended to farmers for diagnosis and treatment of farm problems and to provide some specialist advice to individual farmers.

Farm forestry: Farm Forestry involves practice of forestry in all its aspects in farms and village lands and is usually integrated with other farm operations. It is also known as agro-forestry.

Farm Household System: A group of usually related people who individually or jointly provide management, labour, capital, land and other inputs for production of crops, livestock and also consume at least part of the farm produce.

Farm Yard Manure: It refers to decomposed mixture of dung and urine of farm animals along with the litter and left over material from fodder fed to animals.

Farming System Research: Farming System Research views production unit in farm and household together. It also recognizes inter dependency and inter-relationship between natural human environment so that sustainability can be maintained.

Farming System: Farming System is a decision making and land use unit comprising the farm household cropping and livestock.

Fertilizer: Any natural or manufactured material dry or liquid added to soil in order to supply one or more plant nutrients.

Field Experiment: A field experiment is a research study in a realistic situation in which one or more independent variables are manipulated by the experimenter under as carefully control condition as the situation will permit. The difference between lab experiment and field experiment is that lab experiment has maximum control whereas most field studies operate with less control.

Food security: Can be defined as the "state in which all persons obtain a nutritionally adequate, culturally acceptable diet at all times through local non-emergency sources".

G

Genetic erosion: The loss of genetic diversity in domesticated organisms that have resulted from human reliance on a few genetically uniform varieties of food crop plants and animals.

Green manuring: It is a practice of ploughing or turning into the soil under composed green plant materials for improving the physical condition of the soil.

H

Habitat: The particular environment, characterized by a specific set of environmental conditions, in which a given species occurs

Herbicide: The chemicals used for killing or inhibiting the growth of higher plants.

Heterosis: The production of an exceptionally vigorous and/ or productive hybrid progeny from a directed cross between two pure-breeding plant lines.

Humus: *The* fraction of organic matter in the soil resulting from decomposition and mineralization of organic material.

Hypothesis: Hypothesis is a proposition, condition or principle, which is assumed perhaps without belief in order to draw out the logical consequence and by this method to test fact which are known or may be determined. It is an expected but unconfirmed relationship among two or more variables.

I

Innovation: An innovation is an idea or practice perceived as new by individual in a social system.

Innovativeness: It is the degree to which an individual is relatively earlier in adopting new ideas than other members of his social system.

Integrated Farming Systems (IFS)/Integrated Food and Farming Systems (IFFS): Farming research and policy programs have begun to recognize that by viewing farms and the food production system as an integrated whole, more efficient use can be made of natural, economic, and social resources. Included in this concept are the goals of finding and adopting "integrated and resource-efficient crop and livestock systems that maintain productivity, that are profitable, and that protect the environment and the personal health of farmers and their families," as well as "overcoming the barriers to adoption of more sustainable agricultural systems so these systems can serve as a foundation upon which rural American communities will be revitalized".

Integrated nutrient management: It is a strategy to produce and causably increase soil fertility for sustaining crop productivity through optimum use of organic and inorganic sources.

Integrated Pest Management (IPM): IPM is an ecologically based approach to pest (animal and weed) control that utilizes a multi-disciplinary knowledge of crop/ pest relationships, establishment of acceptable economic thresholds for pest populations and constant field monitoring for potential problems. Management may include such practices as "the use of resistant varieties; crop rotation; cultural practices; optimal use of biological control organisms; certified seed; protective seed treatments; disease-free transplants or rootstock; timeliness of crop cultivation; improved timing of pesticide applications; and removal or 'plow down' of infested plant material". IPM emphasizes a "range of preventive tactics and biological controls to keep pest population within acceptable limits. Reduced risk pesticides are used if other tactics have not been adequately effective, as a last resort and with care to minimize risks".

Integrated Pest Management: Pest control using an array of complementary approaches including natural predators, parasites, pest-resistant varieties, pesticides, and other biological and environmental control practices.

Interaction effect: It is also called snowballing effect. It is a process in which individual in a social system who have adopted the innovation influence those who have not adopted yet acts the community and put these ideas into action.

Inter-cropping: Planting more than one crop in a field using a regular pattern that interleaves each crop in some pattern, a form of polyculture. It is the cultivation of two or more crops simultaneously on the same field with or row arrangement.

Interval Scale: In an interval scale not only the positions are arranged in terms of greater, equal or less but units of interval are also equal.

Interview schedule: Interview schedule refers to a set of questions, which are asked and filled by an interviewer on a face to face situation with another person. A questionnaire refers to a device for securing answer to questions by using a form, which the respondents fill in them.

J

Journal: These are periodical containing magazine information relating to various topics of interest not only for the farmers but also for extension agents.

K

Key informant: The concept of key informant is one among the menu methods of RRA and PRA. Inquiring who are the experts and seeking them out is the method to be adopted for the technique of key informant.

L

Laboratory experiment: Lab experiment is a research study in which the variance of all or nearly all of the possible influential independent variables not pertinent to the problem of investigation is kept at minimum.

Labour productivity: Average outputs produced by labour after the effect of other inputs have been taken into account.

Land use system: The way in which the land is used by a particular group of people with a specified area.

Leadership: Leadership is the activity of influence in which people cooperate to achieve some goal, which they find desirable.

Leaf-area index: *A* measure of leaf covers above a certain area of ground, given by the ratio of total leaf surface area to ground surface area.

Legume: *A* plant in the leguminosae (Fabaceae) family. Most species in this family can fix nitrogen. Any of the legumeinaceae family that may tree, shrubs and herbs many of which have capacity to live in symbiotic relationship with rhizobia.

Ley farming: It is the alteration of food crops with pastures on same piece of land.

Livelihood analysis: Livelihood analysis refers to finding out a degree to which pattern of life differs from one social class to another social class in terms of size of family, type of landholding, annual income, source of income, food habit, expenditure pattern, indebtedness and types of animal owned.

Livestock system: It is a land use unit comprising pasture and herds and auxiliary feed sources transforming plant biomass into animal product.

M

Macronutrient: *A* nutrient plants need in large quantities; the macronutrients include carbon, nitrogen, oxygen, phosphorus, sulfur, and water.

Market oriented Strategy: This is one of the types of rural development strategies gives primary use of market force in shaping rural development.

Mesophyte: A plant that is adapted to environments that are neither very dry nor very wet. Compare with *xerophyte* and *hydrophyte.*

Method of demonstration: It is relatively short time demonstration given before a group to show how to carry out an entirely new practice or old practice in a better way.

Microclimate: The environmental conditions in the immediate vicinity of an organism

Micronutrient: *A* nutrient necessary for plant survival but needed in relatively small quantities.

MIS in Project Management System: Management of information system is used for assessing the performance of development project and is obtained through information system.

Mixed cropping: Growing two or more crops simultaneously with no distinct row arrangement.

Mobility map: It indicates the places where villagers go out of their village for various reasons like agriculture, health, social relation, forestry, animal husbandry, family needs, *etc.*

Model: Model is something which is eminently worth of imitation. Model is the idea to be copied. Simple meaning of model is representatives identically to be copied. A model is essentially recognizable imitation or replica of the original whether workable or not or whether differing or not from the original in size.

Mono-cropping: A crop system in which the population of other than main crop grown in relay, sequential or intercrossing with the main crop are greatly reduced.

Multiple cropping: *The cultivation of two or more crops in succession or with some* overlap in the same field within one year. Double-cropping of rice after wheat is an example. When crops overlap in time, multiple cropping is a form of polyculture.

Multiple cropping: The intensification of cropping in time and space dimensions, growing two or more crops on the same field in a year.

N

Net Product Value (NPV): NPV is one of the economic efficiency indicators which can be used as financial viability indicator. The decision rule would be accept a project if NPV of the project is positive.

Non-Participant Rate: It means interviewer need not actually participate in people's activity but live among them and makes his observation without performing the particular role in the community or organization.

Non-renewable resources: Resources such as oil, coal and mineral host which cannot be regenerated on time scale relevant to human and are exhaustible.

Null Hypothesis: Hypothesis of no relationship among variables called null hypothesis.

O

Off-Campus Training: When training is provided at the convince of the trainees out side the institutions.

On-Campus Training: Residential training or training given at institutions having required infrastructures.

On-Farm Research: It is an approach of adaptive research in farmers field under their management with their perception in view and active participation.

Output: Products or function, which are obtained by means of farming activities and consumed by, farm household or reinvested in farming or may be enhanced or sold.

P

Participatory technology development: It is process of combining the indigenous knowledge and research capacities of local farming communities with that of research and development institution in an interactive way in order to identify generate test and apply new techniques to strengthen the existing technology.

Path-Goal Approach: It indicates most important function of leader to set goals that subordinate to help them to find out the best path for achieving them and remove obstacles.

PERT: Programme Evaluation Review Technique is a method of minimizing production delay, interruption and conflicts of coordinating and synchronizing the various parts of all jobs and expediting the completion of the project.

pH: *The* logarithm of the reciprocal of the concentration of hydrogen ions in a medium, like water or soil (logl0{l/ [H+]}). pH values range from 0 to 14, giving the relative acidity or alkalinity of a medium, with a pH of 7 being neutral, and lower values being acidic, higher values, alkaline.

Phenotype: *The* physical expression of the genotype; an organism's physical characteristics.

Photoperiod: *The* total number of hours of daylight.

Photorespiration: The energetically-wasteful substitution of oxygen for carbon dioxide in the dark reactions of photosynthesis, which occurs when plant stomata close and carbon dioxide concentration declines.

Photosynthate: The simple-sugar end products of photosynthesis.

Plan of Research: It is the overall scheme or programme of research. It includes an outline of what the investigator will do from writing the hypothesis and their operational implications to final analysis.

Polyculture: Cropping systems in which different crop species are grown in mixtures in the same field at the same time.

PRA: Participatory Rural Appraisal is a growing family of approaches and method to enable local people to share enhances and analyzes their knowledge of life and condition to plan and act.

Predator: An animal that consumes other animals to satisfy its nutritive requirements.

Pre-Test: A trial run of an experiment or survey tool with a small number of preliminary subject or respondents to evaluate and rehearse the study procedure and personnel.

Primary Data: Data obtained directly from the original source through a specific investigation rather than from a published source. Social science research depends heavily on primary data collected through observation, schedule interview, *etc.*

Problem Tree: The problem tree is a very useful tool for getting the root of the problem. It provides a frame work for analyzing the dimensions of situation.

Productivity: It can be expressed as output per unit land capital, labour, energy water, nutrients etc.

Project Appraisal: Project appraisal is an exercise which is required before a project is sanctioned. Appraisal means the act of working out a value, quality or condition of the project.

Q

Quota Sampling: It begins by dividing population into relevant strata such as age, gender, geographical locations and some other attributes.

R

Rapid Rural Appraisal: RRA is a systematic semi structured activity carried in the field by a multi-disciplinary team and designed to acquire quickly new information on hypothesis about rural life.

Rapport Building: Rapport building is a key element to learn maximum from informant and other farmers of the village with whom one has to interact.

Relay Cropping: Growing two or more crops simultaneously during the part of life cycle of each. Second crop is planted after the first crop has reached the reproductive stage of growth before it is ready for harvest.

Resource Map: Natural resources to be included in the participatory mapping are soils, trees and forest and water resources.

Result Demonstration: A result demonstration is the method of teaching designed to show by example the practical application of an established fact or group of related facts.

Row Cropping: Growing two or more crops simultaneously where one or more crops are planted in rows.

Rural Development: The building of infrastructure, provision of uncontaminated drinking water education, primary health facility, preventive hygiene and development in agriculture form major components of rural development. It involves a number of projects which are aligned to one another so that they affect various facets of rural economic and social life.

S

Scientific Research: Scientific research is a systematic, empirical and critical investigation of the hypothetical proposition about the presumed relation among natural phenomenon.

Seasonal Calendar: It is often used as to explore constraints and opportunities. Seasonal activities are often called seasonal calendar which indicates month wise, the abnormalities, specialities, threats, problems abundance, shortage with regard to agriculture in a diagrammatical way.

Seed: Seeds may be defined as fertilized ovule consisting an intact embryo, stored food and seed coat which is viable and has got capacity to germinate.

Social Forestry: Social forestry is the practice of forestry for the society by the society specialties for fuel, fodder, fruits and small timber requirements of the nearby surroundings.

Social Marketing: Social marketing is defined as the design, implementation and control of programme calculated to influence the acceptability of official ideas and involving consideration of product, planning, pricing, communication, distribution and marketing research.

Social Research: Social research is a scientific undertaking by means of logical and sympathetic techniques intended to discover new facts or verify or test old facts.

Strategy in Research: It includes methods to gather and analyze data. Strategy implies how research objectives will be reached and how the problems encountered in the research will be tackled.

Structure in Research: The structure of research is more specific and pin pointed. It is the outline, the scheme and the paradigm of the operation of the variables.

Subsidy: The difference between the market price and the price paid by the beneficiary. The difference is paid by the government.

Survey Research: Survey research studies large and small population or universe by selecting and studying samples chosen from the population to discover the relative incidence and interrelation of sociological and psychological variables.

Sustainable Agriculture: It is the successful management of resources for agriculture to satisfy changing human needs while maintaining the enhancing quality of environment and conserving natural resources.

Sustainable Development: Sustainable development must, "meet the needs of the present without compromising the ability of future generations to meet their own needs".

T

Technology Map: It is depict technologies that are found in the villages as well as technology decisions of the farmers. The technology map is an important aspect of PRA.

Time Trend: Time trend or change analysis refers to how to practice and things close to the practice has changed. Changes may be in land use, cropping pattern, migration, fuel used, daily routine diagram. It depicts the fixed and regular way of doing home, farm and outside activities by a farmer or farm women every day in a village from the time a person gets up in the morning from bed till going to bed in the night.

Transfer of Technology: It is defined as a process by which technical know how is disseminated among the members of the society for higher output.

U

Under Development: It is characterized by mass poverty, low level of income, low level of productivity, high level of unemployment.

V

Venn Diagram: The term venn diagram refers to a diagram developed by a person called venn to indicate the contribution or inside and outside agencies and individuals in a decision making process of the inhabitants of a village as perceived by the villagers themselves of the village to indicate the technologies that are adopted, discontinued over adopted, rejected, reinvented for different crops and domestic animals.

W

Weeds: Weeds are the species of plants that grow unwanted or are not useful often prolific and persistent interferes with agricultural operation increase labour cost and reduce crop yield.